Mathematical Analysis 1

theory and solved exercises

Alessio Mangoni

©2020 Alessio Mangoni. All rights reserved.

ISBN: 9798570296077

DR. ALESSIO MANGONI, PHD

Scientist and theoretical particle physicist, researcher on high energy physics and nuclear physics, author of many scientific articles published on international research journals, available at the link:

http://inspirehep.net/author/profile/A.Mangoni.1

https://www.alessiomangoni.it

I edition, November 2020

Contents

Contents 5

Introduction 17

1 Trigonometry 19
1.1 Trigonometric functions 19
1.2 Fundamental relations 31
1.3 Law of sines 32

1.4	Law of cosines	33
1.5	Addition formulas	35
1.6	Prosthaphaeresis formulas	38
1.6.1	Prosthaphaeresis formulas for the sine	39
1.6.2	Prosthaphaeresis formulas for the cosine	40
1.6.3	Prosthaphaeresis formulas for the tangent	41
1.6.4	Prosthaphaeresis formulas for the cotangent	42
1.7	Werner formulas	42
1.7.1	First Werner formula	43
1.7.2	Second Werner formula	43
1.7.3	Third Werner formula	44
1.8	Chord theorem	45
1.9	Area of a generic triangle	49
1.10	Application examples	50
2	**Limits**	**55**
2.1	Introduction	55
2.2	Accumulation point	56
2.3	Definition of limit	56
2.4	Limit from the right and left	57

2.5	Continuity of a function	59
2.6	Uniqueness of the limit	59
2.7	Limit of a sum or product	60
2.8	Theorem of the permanence of the sign	64
2.9	Squeeze theorem	65
2.10	Notable limits	67

3 Sequences and series 73

3.1	Introduction	73
3.2	Definition of sequence	74
3.3	Limit of sequences	75
3.4	Definition of series	76
3.5	Algebraic and geometric sequences	78
3.5.1	Term n-th	78
3.5.2	Partial sum n-th	79
3.5.3	A particular geometric series	81
3.6	Theorems	82
3.7	Comparison test	84
3.8	Asymptotic comparison test	86
3.9	Ratio test	87

3.10	Asymptotic ratio test	88
3.11	Absolute convergence test	89
3.12	Root test	90
3.13	Leibniz's test	91

4 Derivatives ... 95

4.1	Incremental ratio and derivative	95
4.2	Properties of the derivative	97
4.3	Derivatives of elementary functions	99
4.4	Chain rule	102
4.5	Weierstrass theorem	104
4.6	Fermat's theorem on stationary points	104
4.7	Rolle's theorem	105
4.8	Lagrange's theorem	106
4.9	Cauchy's theorem	107
4.10	De L'Hopital's theorem	109
4.11	From mathematics to physics	110

5	**Integrals**	**113**
5.1	Introduction	113
5.2	Definition of integral	116
5.3	Linearity of the integral	117
5.4	Additivity of the integral	117
5.5	Absolute value theorem	117
5.6	Mean value theorem	118
5.7	Fundamental theorem	120
5.8	Primitives of elementary functions	123
5.9	Methods of integration	125
5.9.1	Integration by parts	125
5.9.2	Integration by substitution	126
5.10	From mathematics to physics	127
6	**Exercises**	**131**
6.1	Exercise 1	131
6.2	Exercise 2	132
6.3	Exercise 3	133
6.4	Exercise 4	134

6.5	Exercise 5	135
6.6	Exercise 6	136
6.7	Exercise 7	137
6.8	Exercise 8	137
6.9	Exercise 9	138
6.10	Exercise 10	139
6.11	Exercise 11	139
6.12	Exercise 12	140
6.13	Exercise 13	140
6.14	Exercise 14	141
6.15	Exercise 15	141
6.16	Exercise 16	142
6.17	Exercise 17	142
6.18	Exercise 18	143
6.19	Exercise 19	143
6.20	Exercise 20	144
6.21	Exercise 21	145
6.22	Exercise 22	146
6.23	Exercise 23	147

6.24	Exercise 24	148
6.25	Exercise 25	149
6.26	Exercise 26	150
6.27	Exercise 27	151
6.28	Exercise 28	151
6.29	Exercise 29	152

7 Solutions 153

7.1	Exercise 1	153
7.1.1	Text 153	
7.1.2	Solution 154	
7.2	Exercise 2	155
7.2.1	Text 155	
7.2.2	Solution 155	
7.3	Exercise 3	156
7.3.1	Text 156	
7.3.2	Solution 156	
7.4	Exercise 4	157
7.4.1	Text 157	
7.4.2	Solution 157	

7.5 Exercise 5 — 158
7.5.1 Text — 158
7.5.2 Solution — 159

7.6 Exercise 6 — 160
7.6.1 Text — 160
7.6.2 Solution — 161

7.7 Exercise 7 — 162
7.7.1 Text — 162
7.7.2 Solution — 162

7.8 Exercise 8 — 164
7.8.1 Text — 164
7.8.2 Solution — 164

7.9 Exercise 9 — 165
7.9.1 Text — 165
7.9.2 Solution — 166

7.10 Exercise 10 — 167
7.10.1 Text — 167
7.10.2 Solution — 167

7.11 Exercise 11 — 168
7.11.1 Text — 168
7.11.2 Solution — 169

7.12 Exercise 12 — 169
7.12.1 Text 169
7.12.2 Solution 170

7.13 Exercise 13 — 171
7.13.1 Text 171
7.13.2 Solution 171

7.14 Exercise 14 — 172
7.14.1 Text 172
7.14.2 Solution 173

7.15 Exercise 15 — 174
7.15.1 Text 174
7.15.2 Solution 174

7.16 Exercise 16 — 176
7.16.1 Text 176
7.16.2 Solution 176

7.17 Exercise 17 — 179
7.17.1 Text 179
7.17.2 Solution 179

7.18 Exercise 18 — 180
7.18.1 Text 180
7.18.2 Solution 180

7.19 Exercise 19 — 181
7.19.1 Text — 181
7.19.2 Solution — 181

7.20 Exercise 20 — 182
7.20.1 Text — 182
7.20.2 Solution — 182

7.21 Exercise 21 — 184
7.21.1 Text — 184
7.21.2 Solution — 184

7.22 Exercise 22 — 186
7.22.1 Text — 186
7.22.2 Solution — 186

7.23 Exercise 23 — 187
7.23.1 Text — 187
7.23.2 Solution — 187

7.24 Exercise 24 — 188
7.24.1 Text — 188
7.24.2 Solution — 189

7.25 Exercise 25 — 190
7.25.1 Text — 190
7.25.2 Solution — 190

7.26 Exercise 26   191

7.26.1 Text 191

7.26.2 Solution 191

7.27 Exercise 27   192

7.27.1 Text 192

7.27.2 Solution 192

7.28 Exercise 28   193

7.28.1 Text 193

7.28.2 Solution 193

7.29 Exercise 29   194

7.29.1 Text 194

7.29.2 Solution 195

Introduction

This book on mathematical analysis is intended for both high school and college students to prepare for math exams. The main topics covered are trigonometry, limits, sequences and series, derivatives, integrals. The text contains graphs, figures and examples of application of the theory with various recall to physics. In the second part of the book we propose and solve various original exercises.

1. Trigonometry

1.1 Trigonometric functions

The basic trigonometric functions are the sine and cosine of an angle. Consider the circumference, of unit radius, shown in the figure 1.1.1. Given a point P on the circumference, consider the angle whose measure, in radians, is indicated by x in the figure 1.1.1.

Definition 1.1.1 (sine)**.** We define *sine* of the angle x, denoted

Figure 1.1.1: *Unit circumference with x, sinx e cosx.*

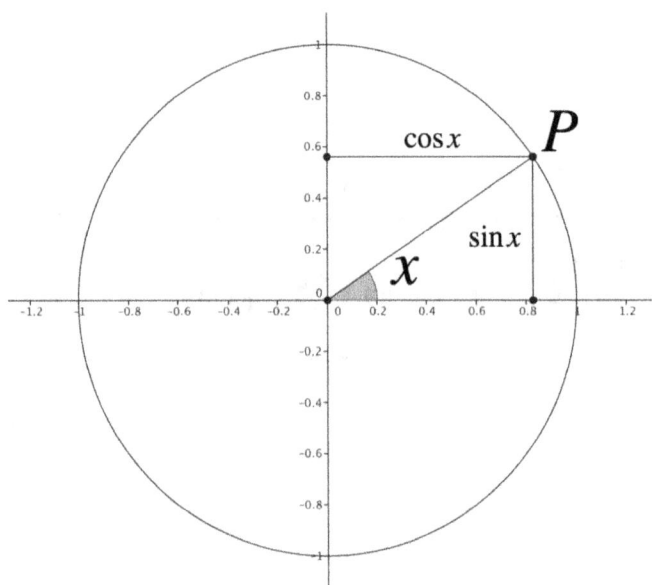

by

$$\sin x,$$

the length, without unit of measure, of the ordinate of the point P. This function, extended to all $\mathbb{R}$, has domain and range respectively

$$D = \mathbb{R}, \quad C = [-1, 1]$$

and is periodic of period $T = 2\pi$. Figure 1.1.2 shows the plot of the function.

1.1 Trigonometric functions

Figure 1.1.2: *Plot of the function* sin*x*.

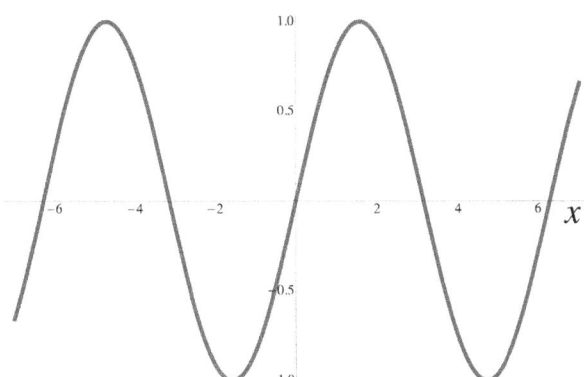

Similarly we have the following

Definition 1.1.2 (cosine). We define *cosine* of the angle *x*, and denoted by

$$\cos x,$$

the length, always without unit of measure, of the abscissa of the point *P*. This function has, like sin*x*, domain and range

$$D = \mathbb{R}, \qquad C = [-1, 1]$$

and is periodic of period $T = 2\pi$. In Figure 1.1.3 is shown the plot of the function.

We observe that the plots of the sine and cosine functions are one coincident with the other translated by an amount of $\pi/2$,

Figure 1.1.3: *Plot of the function* cos *x.*

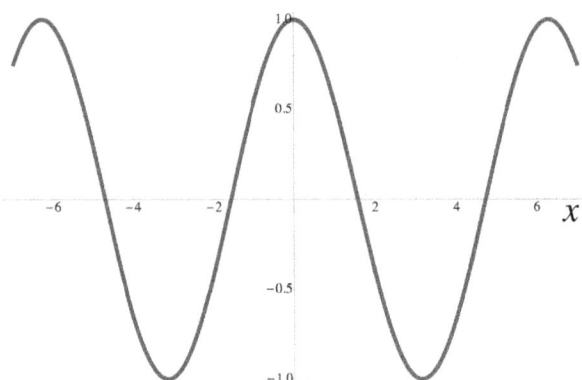

i.e.

$$\sin(x) = \cos\left(x - \frac{\pi}{2}\right).$$

If the circumference in Figure 1.1.1 were of not unitary radius R, since the angle x is invariant, the abscissa and the ordinate of the point P would be $R\cos x$ and $R\sin x$ respectively.

We now define the tangent of the angle x.

Definition 1.1.3 (tangent). The *tangent* of the angle x is defined as follows

$$\tan x := \frac{\sin x}{\cos x}.$$

1.1 Trigonometric functions

Figure 1.1.4: *Plot of the function* $\tan x$.

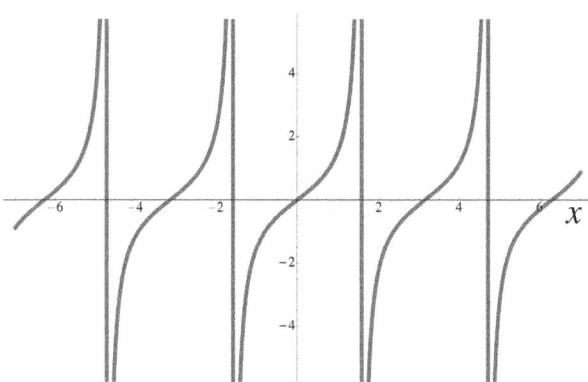

This function has domain and range

$$D = \mathbb{R} \setminus \left\{\frac{\pi}{2} + k\pi\right\}, \forall k \in \mathbb{Z}, \qquad C = \mathbb{R}$$

due to zeroes in the denominator and is periodic of period $T = \pi$. In Figure 1.1.4 is shown the plot of the function.

There are also other functions, of secondary use, related to those introduced so far.

Definition 1.1.4 (secant)**.** The *secant* of the angle x is defined as

$$\sec x := \frac{1}{\cos x}.$$

This function has domain and range

$$D = \mathbb{R} \setminus \left\{\frac{\pi}{2} + k\pi\right\}, \forall k \in \mathbb{Z}, \qquad C = \mathbb{R}$$

Figure 1.1.5: *Plot of the function* sec*x*.

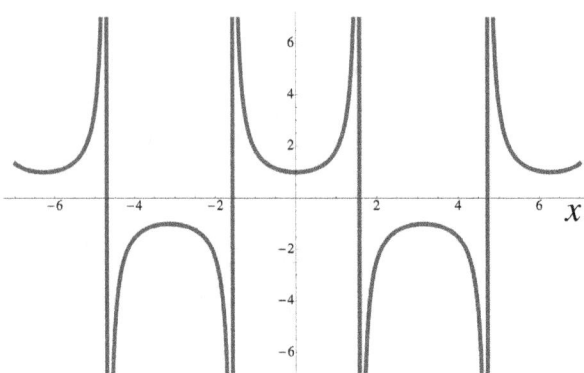

and is periodic of period $T = 2\pi$. In Figure 1.1.5 is shown the plot of the function.

Definition 1.1.5 (cosecant). The *cosecant* of the angle x is defined as
$$\csc x := \frac{1}{\sin x}.$$
This function has domain and range
$$D = \mathbb{R} \setminus \{k\pi\}, \forall k \in \mathbb{Z}, \qquad C = \mathbb{R}$$
and is periodic of period $T = 2\pi$. In Figure 1.1.6 is shown the plot of the function.

Definition 1.1.6 (cotangent). The *cotangent* of the angle x is

1.1 Trigonometric functions

Figure 1.1.6: *Plot of the function* csc*x*.

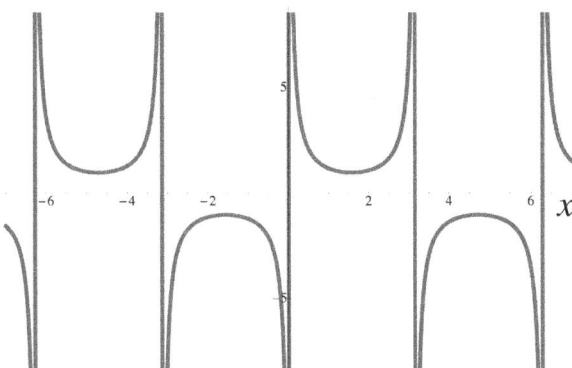

defined as
$$\cot x := \frac{1}{\tan x} = \frac{\cos x}{\sin x}.$$

This function has domain and range

$$D = \mathbb{R} \setminus \{k\pi\}, \forall k \in \mathbb{Z}, \qquad C = \mathbb{R}$$

and is periodic of period $T = \pi$. In Figure 1.1.7 is shown the plot of the function.

It is also possible to define the inverse functions of all these trigonometric functions, however we must restrict their domain to make them bijective, that is a necessary condition for inverting them. Usually the domain is restricted to the ranges $[-\pi/2, \pi/2]$ or $[0, \pi]$. We now define the inverse functions of

Figure 1.1.7: *Plot of the function* cot*x*.

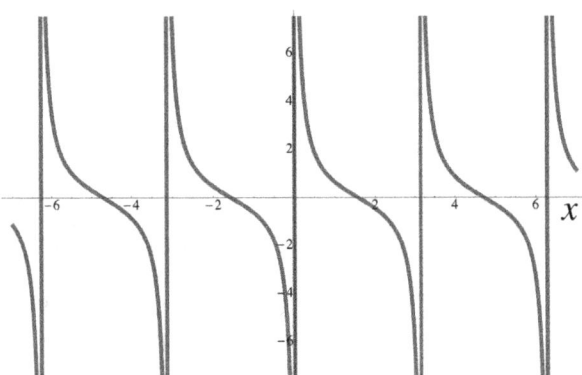

sin*x*, cos*x* and tan*x* which are called arccosine, arccosine and arctangent.

Definition 1.1.7 (arccosine). We define *arcsine* of $x \in [-1, 1]$, denoted by

$$\arcsin x,$$

the angle (unique thanks to the restriction) in the range

$$\left[-\frac{\pi}{2}, \frac{\pi}{2}\right]$$

such that

$$\sin(\arcsin x) = x.$$

This function is the inverse of sin*x* and has domain and range

$$D = [-1, 1], \qquad C = \left[-\frac{\pi}{2}, \frac{\pi}{2}\right].$$

1.1 Trigonometric functions

Figure 1.1.8: *Plot of the function* arcsin *x*.

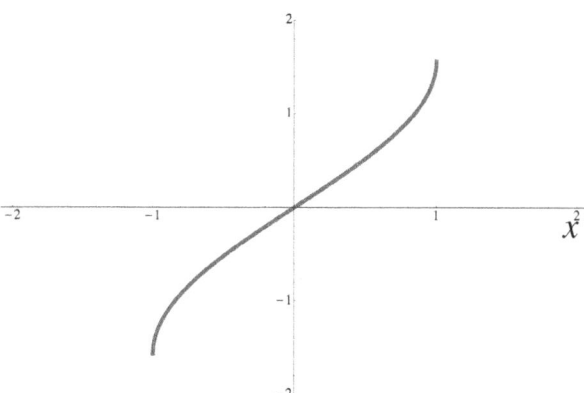

In Figure 1.1.8 is shown the plot of the function.

Definition 1.1.8 (arccosine). We define *arccosine* of $x \in [-1, 1]$, denoted by

$$\arccos x,$$

the angle (unique thanks to the restriction) in the range

$$[0, \pi]$$

such that

$$\cos(\arccos x) = x.$$

This function is the inverse of $\cos x$ and has domain and range

$$D = [-1, 1], \qquad C = [0, \pi].$$

Figure 1.1.9: *Plot of the function* arccos*x*.

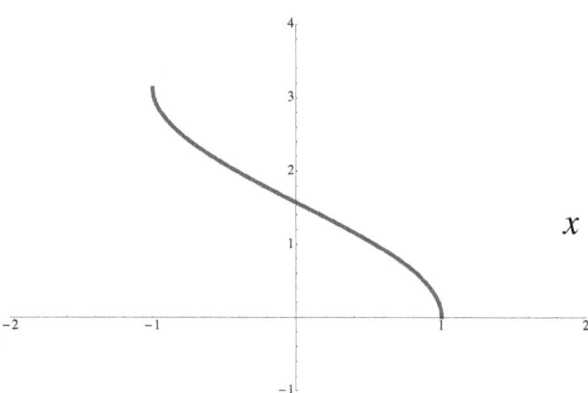

In Figure 1.1.9 is shown the plot of the function.

Definition 1.1.9 (arctangent). We define *arctangent* of x, denoted by

$$\arctan x,$$

the angle (unique thanks to the restriction) in the range

$$\left(-\frac{\pi}{2}, \frac{\pi}{2}\right)$$

such that

$$\tan(\arctan x) = x.$$

This function is the inverse of $\tan x$ and has domain and range

$$D = \mathbb{R}, \qquad C = \left(-\frac{\pi}{2}, \frac{\pi}{2}\right).$$

1.1 Trigonometric functions

Figure 1.1.10: *Plot of the function* arctan *x.*

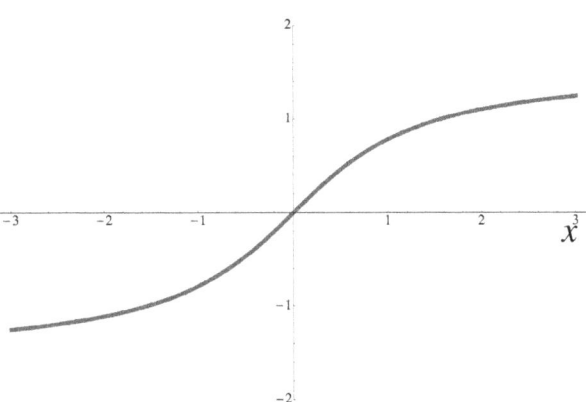

In Figure 1.1.10 is shown the plot of the function.

We now define the inverse functions of sec x and csc x, called arcsecant and arccosecant, respectively.

Definition 1.1.10 (arcsecant)**.** We define *arcsecant* of $x \notin (-1, 1)$, denoted by

$$\operatorname{arcsec} x,$$

the angle (unique thanks to the restriction) in the range

$$[0, \pi]$$

such that

$$\sec(\operatorname{arcsec} x) = x.$$

Figure 1.1.11: *Plot of the function* arcsec*x*.

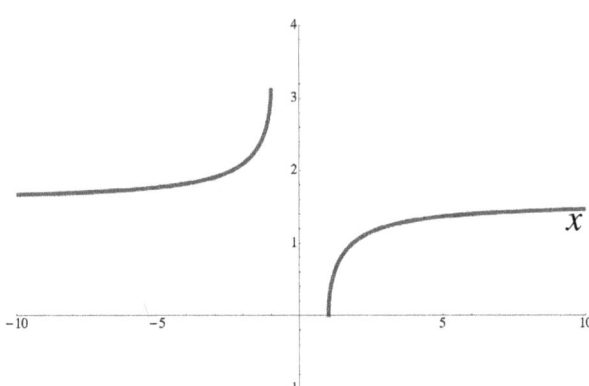

This function is the inverse of sec*x* and has domain and range

$$D = \mathbb{R} \setminus (-1, 1), \qquad C = [0, \pi].$$

In Figure 1.1.11 is shown the plot of the function.

Definition 1.1.11 (arccosecant). We define *arccosecant* of $x \notin (-1, 1)$, denoted by

$$\text{arccsc}\, x,$$

the angle (unique thanks to the restriction) in the range

$$\left[-\frac{\pi}{2}, \frac{\pi}{2} \right]$$

such that

$$\csc(\text{arccsc}\, x) = x.$$

1.2 Fundamental relations

Figure 1.1.12: *Plot of the function* arccsc x.

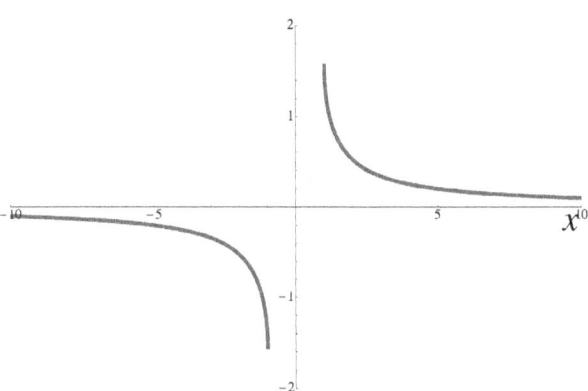

This function is the inverse of csc$x$ and has domain and range

$$D = \mathbb{R} \setminus (-1, 1), \qquad C = \left[-\frac{\pi}{2}, \frac{\pi}{2}\right].$$

In Figure 1.1.12 is shown the plot of the function.

1.2 Fundamental relations

Applying the Pythagorean theorem to the triangle of Figure 1.1.1 having for catheti cos$x$ and sin$x$ and for hypotenuse 1 we immediately obtain the fundamental trigonometric identity

$$\sin^2 x + \cos^2 x = 1,$$

from which we can obtain also

$$\sin x = \pm\sqrt{1 - \cos^2 x},$$

$$\cos x = \pm\sqrt{1-\sin^2 x}.$$

Proposition 1.2.1 (relations in a right triangle). *As we have seen, in a right triangle having the two catheti of length a and b and the hypotenuse of length c, indicated by α and β the opposite angles respectively to a and b, we can write*

$$a = c\cos\beta = c\sin\alpha,$$

$$b = c\cos\alpha = c\sin\beta,$$

$$\frac{a}{b} = \tan\alpha = \cot\beta.$$

1.3 Law of sines

Theorem 1.3.1 (law of sines). *Given any triangle, denoted by a, b and c the lengths of the sides and by α, β and γ the respective opposite angles, we have*

$$\frac{a}{\sin\alpha} = \frac{b}{\sin\beta} = \frac{c}{\sin\gamma}. \qquad (1.3.1)$$

Proof

With reference to the Figure 1.7.1, which shows any triangle with the height relative to the side c plotted, we can apply

1.4 Law of cosines

the trigonometric relations in the two right triangles that have been formed, obtaining

$$h = b \sin \alpha,$$

$$h = a \sin \beta$$

and equaling h

$$b \sin \alpha = a \sin \beta,$$

hence the first part of Eq. (1.3.1)

$$\frac{a}{\sin \alpha} = \frac{b}{\sin \beta}.$$

For the second part, proceed in the same way by tracing another of the two heights relative to the remaining sides b and c.

1.4 Law of cosines

Theorem 1.4.1 (law of cosines). *Given any triangle, denoted by a, b and c the lengths of the sides and by α, β and γ the respective opposite angles, we have*

$$b^2 = a^2 + c^2 - 2ac\cos\beta,$$

Chapter 1. Trigonometry

$$a^2 = b^2 + c^2 - 2bc\cos\alpha,$$

$$c^2 = a^2 + b^2 - 2ab\cos\gamma.$$

Proof

Always referring to the Figure 1.7.1, called c_1 and c_2 the two parts into which the side c is divided by the height h, the Pythagorean theorem can be applied to the two right triangles to obtain

$$h^2 + c_1^2 = a^2, \qquad (1.4.1)$$

$$h^2 + c_2^2 = b^2.$$

Moreover, since $c_2 = c - c_1$, the last formula becomes

$$h^2 + c^2 + c_1^2 - 2cc_1 = b^2$$

and, using Eq. (1.4.1),

$$c^2 + a^2 - 2cc_1 = b^2.$$

Using the trigonometric relations on the right triangle with sides (a, h, c_1) it can be observed that

$$c_1 = a\cos\beta$$

and, finally, we arrive at the first of the relations of the theorem

$$c^2 + a^2 - 2ac\cos\beta = b^2.$$

The other two are obtained immediately, given the arbitrariness of the choice of sides in the proof.

1.5 Addition formulas

It can be shown that the following addition formulas hold for trigonometric functions. Given any two angles x and y, we have

$$\sin(x \pm y) = \sin x \cos y \pm \sin y \cos x, \qquad (1.5.1)$$

$$\cos(x \pm y) = \cos x \cos y \mp \sin x \sin y, \qquad (1.5.2)$$

$$\tan(x \pm y) = \frac{\tan x \pm \tan y}{1 - \tan x \tan y}.$$

We observe that to prove these three relations it is enough to prove that the first is an identity, in fact by deriving it with respect to x or y we obtain the second and from their combination the third. So we want to prove the first relation of

Eq.(1.5.1). For this proof we will use the law of sines and the law of cosines which are shown below. Consider any triangle with sides a, b, c and angles α, β, γ as shown in Figure 1.7.1. We want to prove that

$$\sin(\alpha + \beta) = \sin\alpha \cos\beta + \sin\beta \cos\alpha.$$

Thanks to the laws of sines and cosines we can write

$$\cos\alpha = \frac{b^2 + c^2 - a^2}{2bc},$$

$$\cos\beta = \frac{a^2 + c^2 - b^2}{2ac},$$

$$\sin\alpha = \frac{a}{c}\sin\gamma,$$

$$\sin\beta = \frac{b}{c}\sin\gamma.$$

Meanwhile, we note that being a triangle

$$\alpha + \beta = \pi - \gamma$$

and hence

$$\sin(\alpha + \beta) = \sin(\pi - \gamma) = \sin\gamma,$$

so we have just to prove that

$$\sin\gamma = \sin\alpha \cos\beta + \sin\beta \cos\alpha.$$

1.5 Addition formulas

Thanks to the relations written above we can write

$$\sin\alpha\cos\beta = \frac{a}{c}\frac{a^2+c^2-b^2}{2ac}\sin\gamma,$$

$$\sin\beta\cos\alpha = \frac{b}{c}\frac{b^2+c^2-a^2}{2bc}\sin\gamma,$$

from which

$$\sin\alpha\cos\beta + \sin\beta\cos\alpha = \frac{\sin\gamma}{2c^2}(a^2+c^2-b^2+b^2$$
$$+ c^2-a^2) = \frac{\sin\gamma}{2c^2}(2c^2) = \sin\gamma,$$

which concludes the proof. The first addition formula has been proved only for two arbitrary angles whose sum is less than π. However, it can be noted that by placing

$$\alpha' = \pi - \alpha,$$

$$\beta' = \pi - \beta,$$

$$\gamma' = \pi - \gamma,$$

α' and β' are two arbitrary angles whose sum is greater than π. It can be seen that the addition formula in Eq. (1.5.1) applies also to these angles

$$\sin(\alpha'+\beta') = \sin\alpha'\cos\beta' + \sin\beta'\cos\alpha',$$

in fact

$$\sin(\alpha'+\beta') = \sin(2\pi - (\alpha+\beta)) = \sin(\alpha+\beta),$$

$$\begin{aligned}\sin\alpha'\cos\beta' &= \sin(\pi+\alpha)\cos(\pi+\beta)\\ &= (-\sin\alpha)(-\cos\beta)\\ &= \sin\alpha\cos\beta,\end{aligned}$$

$$\begin{aligned}\sin\beta'\cos\alpha' &= \sin(\pi+\beta)\cos(\pi+\alpha)\\ &= (-\sin\beta)(-\cos\alpha)\\ &= \sin\beta\cos\alpha.\end{aligned}$$

We have therefore demonstrated the validity of Eq. (1.5.1) both for two arbitrary angles whose sum is less than π and for two whose sum is greater than π. It remains only the case in which the sum is π, in this case

$$\begin{aligned}\sin\pi = 0 &= \sin\alpha\cos(\pi-\alpha) + \sin(\pi-\alpha)\cos\alpha\\ &= -\sin\alpha\cos\alpha + \sin\alpha\cos\alpha = 0\end{aligned}$$

and this concludes the proof.

1.6 Prosthaphaeresis formulas

The following are called Prosthaphaeresis formulas

1.6 Prosthaphaeresis formulas

1.6.1 Prosthaphaeresis formulas for the sine

$$\sin x \pm \sin y = 2\sin\left(\frac{x \pm y}{2}\right)\cos\left(\frac{x \mp y}{2}\right), \qquad (1.6.1)$$

Proof

We can write

$$\sin x \pm \sin y = \sin\left(\frac{x+y}{2} + \frac{x-y}{2}\right) \pm \sin\left(\frac{x+y}{2} + \frac{y-x}{2}\right),$$

using the Eq. (1.5.1),

$$\begin{aligned}
\sin x \pm \sin y &= \sin\left(\frac{x+y}{2}\right)\cos\left(\frac{x-y}{2}\right) \\
&+ \sin\left(\frac{x-y}{2}\right)\cos\left(\frac{x+y}{2}\right) \\
&\pm \sin\left(\frac{x+y}{2}\right)\cos\left(\frac{y-x}{2}\right) \\
&\pm \sin\left(\frac{y-x}{2}\right)\cos\left(\frac{x+y}{2}\right),
\end{aligned}$$

knowing that

$$\cos\left(\frac{y-x}{2}\right) = \cos\left(\frac{x-y}{2}\right),$$

$$\sin\left(\frac{y-x}{2}\right) = -\sin\left(\frac{x-y}{2}\right),$$

we obtain

$$\sin x \pm \sin y = \sin\left(\frac{x+y}{2}\right)\cos\left(\frac{x-y}{2}\right)$$
$$+ \sin\left(\frac{x-y}{2}\right)\cos\left(\frac{x+y}{2}\right)$$
$$\pm \sin\left(\frac{x+y}{2}\right)\cos\left(\frac{x-y}{2}\right)$$
$$\mp \sin\left(\frac{x-y}{2}\right)\cos\left(\frac{x+y}{2}\right),$$

$$\sin x \pm \sin y = \sin\left(\frac{x+y}{2}\right)\cos\left(\frac{x-y}{2}\right)(1\pm 1)$$
$$+ \sin\left(\frac{x-y}{2}\right)\cos\left(\frac{x+y}{2}\right)(1\mp 1).$$

Depending on whether the sign is $+$ or $-$ we have

$$\sin x + \sin y = 2\sin\left(\frac{x+y}{2}\right)\cos\left(\frac{x-y}{2}\right),$$

$$\sin x - \sin y = 2\sin\left(\frac{x-y}{2}\right)\cos\left(\frac{x+y}{2}\right)$$

and this concludes the proof.

1.6.2 Prosthaphaeresis formulas for the cosine

$$\cos x + \cos y = 2\cos\left(\frac{x+y}{2}\right)\cos\left(\frac{x-y}{2}\right). \quad (1.6.2)$$

$$\cos x - \cos y = -2\sin\left(\frac{x+y}{2}\right)\sin\left(\frac{x-y}{2}\right). \quad (1.6.3)$$

1.6 Prosthaphaeresis formulas

Proof

To prove these two formulas we observe that, being the Eq. (1.6.1) an identity, we can derive with respect to x obtaining

$$\cos x = \cos\left(\frac{x\pm y}{2}\right)\cos\left(\frac{x\mp y}{2}\right)$$
$$- \sin\left(\frac{x\pm y}{2}\right)\sin\left(\frac{x\mp y}{2}\right)$$

and with respect to y, obtaining

$$\pm\cos y = \pm\cos\left(\frac{x\pm y}{2}\right)\cos\left(\frac{x\mp y}{2}\right)$$
$$\pm \sin\left(\frac{x\pm y}{2}\right)\sin\left(\frac{x\mp y}{2}\right),$$

from which, adding,

$$\cos x \pm \cos y = \cos\left(\frac{x\pm y}{2}\right)\cos\left(\frac{x\mp y}{2}\right)(1\pm 1)$$
$$+ \sin\left(\frac{x\pm y}{2}\right)\sin\left(\frac{x\mp y}{2}\right)(-1\pm 1),$$

that are the Eqs. (1.6.2) and (1.6.3).

1.6.3 Prosthaphaeresis formulas for the tangent

Below we show the Prosthaphaeresis formulas for tangent and cotagent.

$$\tan x \pm \tan y = \frac{\sin(x\pm y)}{\cos x \cos y} \quad \forall x, y \in \mathbb{R}\setminus\left\{\frac{\pi}{2}+k\pi\right\}, \forall k \in \mathbb{Z}.$$

Proof

We can write

$$\tan x \pm \tan y = \frac{\sin x}{\cos x} \pm \frac{\sin y}{\cos y}$$

$$= \frac{\sin x \cos y \pm \sin y \cos x}{\cos x \cos y}$$

$$= \frac{\sin(x \pm y)}{\cos x \cos y},$$

where the last step follows from the addition formula shown in Eq. (1.5.1).

1.6.4 Prosthaphaeresis formulas for the cotangent

$$\cot x \pm \cot y = \frac{\sin(y \pm x)}{\sin x \sin y} \quad \forall x, y \in \mathbb{R} \setminus \{k\pi\}, \forall k \in \mathbb{Z}.$$

Proof

Identically to the previous proof

$$\cot x \pm \cot y = \frac{\cos x}{\sin x} \pm \frac{\cos y}{\sin y}$$

$$= \frac{\sin y \cos x \pm \sin x \cos y}{\sin x \sin y}$$

$$= \frac{\sin(y \pm x)}{\sin x \sin y}.$$

1.7 Werner formulas

The following are called Werner formulas

1.7 Werner formulas

1.7.1 First Werner formula

$$\sin x \sin y = \frac{1}{2}\left(\cos(x-y) - \cos(x+y)\right).$$

Proof

The relation can be written, thanks to Eq. (1.5.2), as

$$\begin{aligned}\sin x \sin y &= \frac{1}{2}(\cos x \cos y + \sin x \sin y \\ &\quad - \cos x \cos y + \sin x \sin y) \\ &= \frac{1}{2}\left(\sin x \sin y + \sin x \sin y\right) \\ &= \frac{1}{2}(2\sin x \sin y),\end{aligned}$$

which is an identity.

1.7.2 Second Werner formula

$$\sin x \cos y = \frac{1}{2}\left(\sin(x+y) + \sin(x-y)\right),$$

Proof

The report can be written, as before,

$$\begin{aligned}\sin x \cos y &= \frac{1}{2}\left(\sin x \cos y + \cos x \sin y + \sin x \cos y - \cos x \sin y\right), \\ &= \frac{1}{2}\left(\sin x \cos y + \sin x \cos y\right) \\ &= \frac{1}{2}(2\sin x \cos y),\end{aligned}$$

which is an identity.

1.7.3 Third Werner formula

$$\cos x \cos y = \frac{1}{2}\left(\cos(x+y) + \cos(x-y)\right).$$

Proof

We can write, as done before,

$$\begin{aligned}\cos x \cos y &= \frac{1}{2}\left(\cos x \cos y - \sin x \sin y\right. \\ &\quad \left. + \cos x \cos y + \sin x \sin y\right) \\ &= \frac{1}{2}\left(\cos x \cos y + \cos x \cos y\right) \\ &= \frac{1}{2}(2\cos x \cos y),\end{aligned}$$

which is an identity.

Figure 1.7.1: *A generic triangle with the height h relative to the side c.*

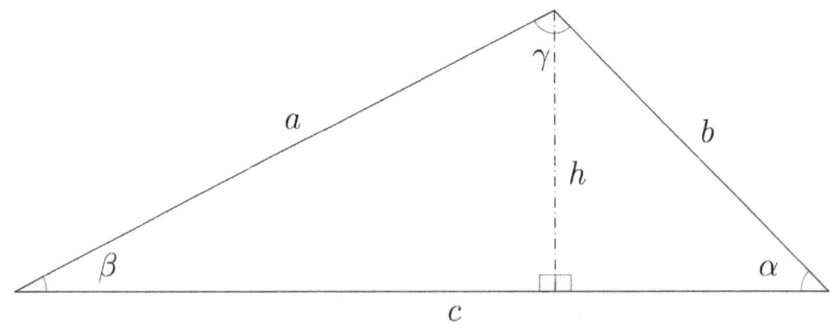

1.8 Chord theorem

Before showing the chord theorem it is necessary to give the following

Definition 1.8.1 (angle at the centre). Given a circumference and one of its chords, the angle subtended by the chord with the vertex at the center of the circumference is called *angle at the centre*.

Definition 1.8.2 (angle at the circumference). Given a circumference and one of its chords, the angle subtended by the chord with the vertex on any point of the circumference is called *angle at the circumference*.

Proposition 1.8.1 (relation between angle at the centre and angle at circumference). *Given a circumference and one of its chords, let α and β be the angle at the centre and the angle at the circumference subtended by the chord, respectively. Then we have*
$$\alpha = 2\beta.$$

Proof

With reference to the Figure 1.8.1 let $\mathcal{C}$ be the chosen chord and α and β the angles at the centre and the angle at the circumference, respectively. In the Figure are shown three rays which identify three isosceles triangles with angles δ, λ and γ. Since the sum of the interior angles of any triangle is π radians we have

$$2(\lambda + \gamma + \delta) = \pi,$$

$$\lambda = \frac{\pi - \alpha}{2}.$$

For construction

$$\beta = \gamma + \delta$$

and therefore we obtain

$$2\left(\frac{\pi - \alpha}{2} + \beta\right) = \pi,$$

from which, finally,

$$\alpha = 2\beta.$$

Note that from this proof it follows that the amplitude of the angle at the circumference is the same regardless of the chosen point.

Theorem 1.8.2 (chord theorem). *In a circle of radius r consider a chord of length c, which subtends an angle at the center, α, and one at the circumference, β. The theorem states*

1.8 Chord theorem

Figure 1.8.1: *Circumference with a chosen chord $\mathcal{C}$ and the two angles at the center and at the circumference, α and β.*

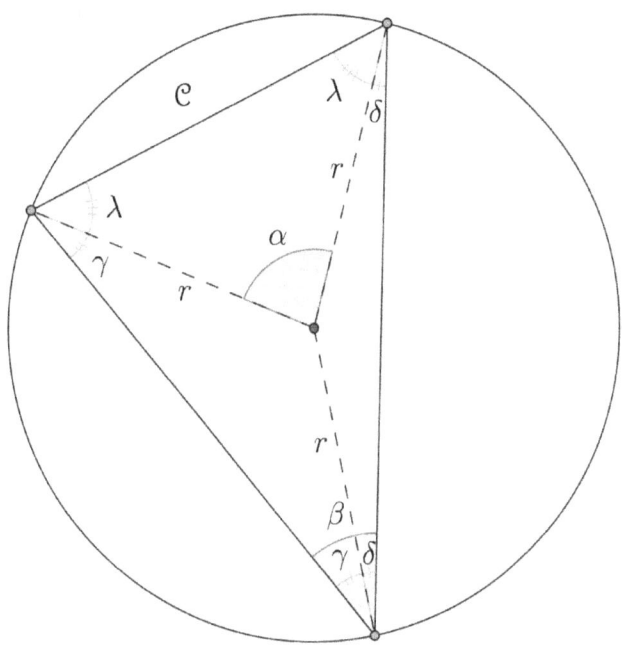

that

$$c = 2r\sin\beta = 2r\sin\frac{\alpha}{2}.$$

Proof

Using the sine theorem shown previously, with the choice of the angle at the circumference according to Figure 1.8.2, we can write the relation

$$\frac{c}{\sin \alpha} = \frac{2r}{\sin \frac{\pi}{2}} = 2r,$$

in fact, the largest triangle shown in the figure is a right triangle because it is inscribed in a semi-circumference. This fact can be proved by observing that the angle that is supposed to be right is given by the sum of the two angles β and δ. In fact, the base angles of the isosceles triangle with sides r, r, c are δ and δ, then

$$\delta = \frac{\pi - \alpha}{2},$$

moreover from the proposition on the angles at the center and at the circumference we have

$$\alpha = 2\beta$$

and therefore

$$\delta = \frac{\pi}{2} - \beta.$$

Finally the angle is right because

$$\delta + \beta = \frac{\pi}{2}.$$

1.9 Area of a generic triangle

Figure 1.8.2: *Circumference with a chosen chord c and the two angles at the center and at the circumference, α and β.*

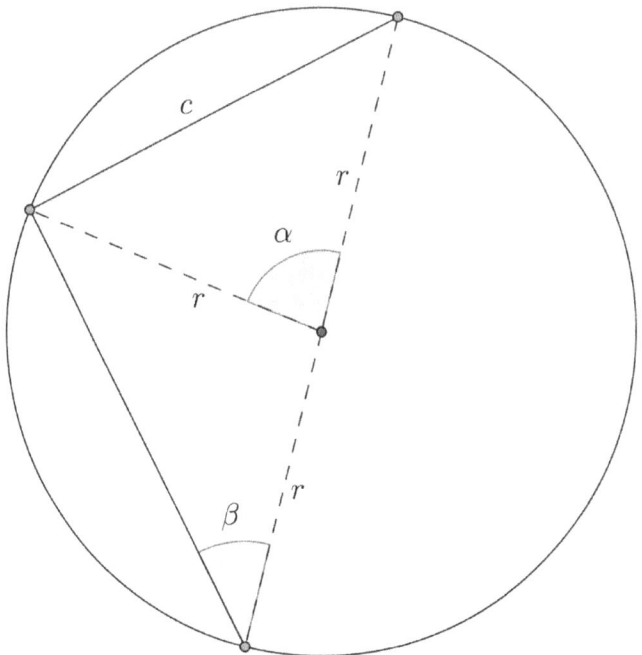

1.9 Area of a generic triangle

Theorem 1.9.1 (area of a generic triangle). *Given a generic triangle, denoted with a, b and c the lengths of its sides and with α, β and γ the respective opposite angles, we can calcu-*

late the area A of the triangle using the following expressions

$$A = \frac{1}{2}ab \sin \gamma,$$

$$A = \frac{1}{2}bc \sin \alpha,$$

$$A = \frac{1}{2}ac \sin \beta.$$

Proof

With reference to the Figure 1.7.1 the area is

$$A = \frac{1}{2}ch.$$

Using the trigonometric relations for h we have

$$h = a \sin \beta,$$

and therefore

$$A = \frac{1}{2}ac \sin \beta.$$

Similarly for the other sides, in fact their choice in the figure is arbitrary.

1.10 Application examples

As a first example of trigonometry application, consider the Figure 1.10.1, where is shown a tower. Suppose we want to

1.10 Application examples

Figure 1.10.1: *Tower of height h which can be seen as the side of a right triangle.*

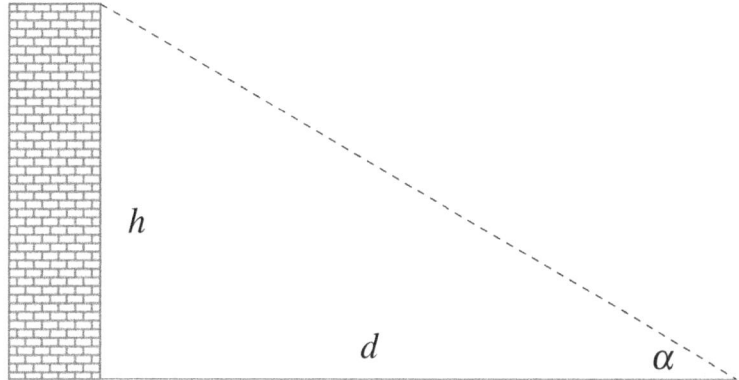

estimate its height, because we cannot measure it directly. Using the formulas obtained previously we have

$$h = d \tan \alpha.$$

So we need only to measure the horizontal distance d and estimate the angle α to obtain the height.

As a second example, suppose we want to estimate the distance of a star that is quite distant and visible from the Earth. Figure 1.10.2 shows the Sun in the center with the Earth in two opposite positions that differ by a temporal distance of

Figure 1.10.2: *Calculation of distances with the parallax method.*

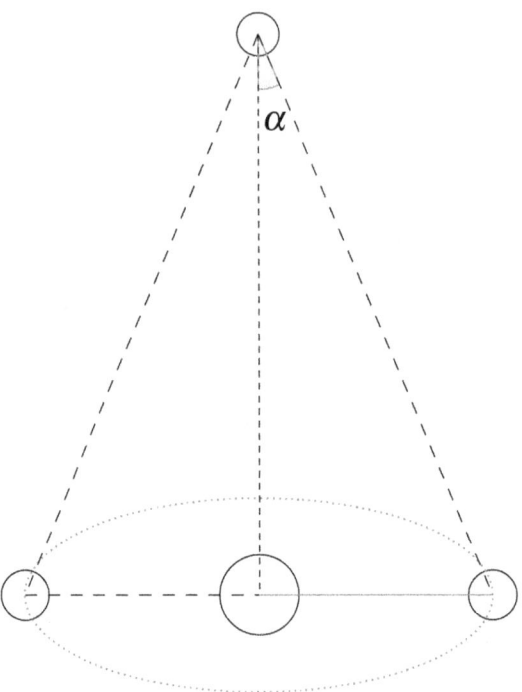

six months. By measuring the angle α, shown in the figure, it is possible to estimate the distance of the star knowing that the distance d between the Earth and the Sun is, on average, 1 astronomical unit, or $1.49 \cdot 10^{11}$ meters. Therefore, in astronomical units ($U.A.$), the distance D of the star is

$$D = \frac{1}{\sin \alpha} \approx \frac{1}{\alpha},$$

since $\sin \alpha \approx \alpha$ for small angles.

2. Limits

2.1 Introduction

The limit of a function $f(x)$ indicates, intuitively, the value the function tends to as its variable x approaches a certain chosen value x_0, not necessarily belonging to the domain of $f(x)$.

2.2 Accumulation point

In order to give a rigorous definition of limit it is necessary to introduce the concept of accumulation point through the following:

Definition 2.2.1 (accumulation point). Given a set $D \subseteq \mathbb{R}$, any point $x_0 \in \mathbb{R}$ is said to be an accumulation point for D if

$$\forall \delta > 0 \ \exists d \in D : d \in (x_0 - \delta, x_0 + \delta),$$

namely, taking any neighborhood of x_0 there is at least one element of this neighborhood that also belongs to the D set.

2.3 Definition of limit

Now that the concept of an accumulation point for a set has been introduced, a rigorous definition of a limit can be given.

Definition 2.3.1 (limit of a function). Given a function $f(x)$ and an accumulation point x_0 for its domain D, l is said to be the limit of f for x that approaches x_0 and it is indicates with

$$\lim_{x \to x_0} f(x) = l$$

if
$$\forall \varepsilon > 0 \; \exists \delta > 0 : \forall x \in (x_0 - \delta, x_0 + \delta) \cap D$$
$$\implies f(x) \in (l - \varepsilon, l + \varepsilon),$$

otherwise, denoting with $I(x_0, \delta)$ a neighborhood of x_0 of radius δ,

$$\forall \varepsilon > 0 \; \exists \delta > 0 : \forall x \in I(x_0, \delta) \cap D \implies f(x) \in I(l, \varepsilon).$$

Finally, a completely equivalent way of defining the limit is the following

$$\forall \varepsilon > 0 \; \exists \delta > 0 : \forall x \in D : |x - x_0| < \delta$$
$$\implies |f(x) - l| < \varepsilon.$$

From now on we will assume that x belongs to the function domain when we write $f(x)$.

2.4 Limit from the right and left

Starting from the definition of limit, we can define also the limit from the right and the limit from the left. We give the following

Definition 2.4.1 (limit from the right). Under the same conditions as the definition of limit, l is said to be the limit from the right of f for x tending towards x_0 and is denoted by

$$\lim_{x \to x_0^+} f(x) = l$$

if

$$\forall \varepsilon > 0 \ \exists \delta > 0 : \forall x \in (x_0, x_0 + \delta) \implies f(x) \in (l - \varepsilon, l + \varepsilon).$$

Similarly:

Definition 2.4.2 (limit from the left). We say that l is the limit from the left of f for x tending towards x_0 and denoted is by

$$\lim_{x \to x_0^-} f(x) = l,$$

if

$$\forall \varepsilon > 0 \ \exists \delta > 0 : \forall x \in (x_0 - \delta, x_0) \implies f(x) \in (l - \varepsilon, l + \varepsilon).$$

If the limits of a function from the left and right exist and are equal we will say that the function admits the limit (in the general sense) and it equals their common value.

2.5 Continuity of a function

Definition 2.5.1 (Continuity). A function $f(x)$ is said to be continuous at a point x_0, belonging to its domain D and being an accumulation point for D, if

$$\lim_{x \to x_0^+} f(x) = \lim_{x \to x_0^-} f(x) = f(x_0),$$

in other words, if the limits from the right and left for x that tends to x_0 exist finites, they coincide and are equal to the value that the function assumes at that point. In general, a function is said to be continuous if it is continuous in every point of its domain.

2.6 Uniqueness of the limit

Theorem 2.6.1 (uniqueness theorem for limits). *If the limit of a function $f(x)$ exists for $x \to x_0$ then it is unique.*

Proof

Let us suppose, by absurd, that there are two different limits l and l', with

$$l \neq l' \tag{2.6.1}$$

then, for the definition of limit, there will be an appropriate neighborhood of x_0 so for each x of this neighborhood we can write[1],

$$\begin{cases} l - \varepsilon < f(x) < l + \varepsilon \\ l' - \varepsilon < f(x) < l' + \varepsilon \end{cases},$$

with arbitrary $\varepsilon > 0$. From which

$$\begin{cases} l - \varepsilon < f(x) < l' + \varepsilon \\ l' - \varepsilon < f(x) < l + \varepsilon \end{cases},$$

or

$$\begin{cases} l - l' > 2\varepsilon \\ l - l' < 2\varepsilon \end{cases},$$

which lead to an absurdity. We must therefore conclude that $l = l'$ and hence that the limit is unique.

2.7 Limit of a sum or product

Proposition 2.7.1 (limit of a sum). *The limit, for $x \to x_0$, of the sum of two functions is equal to the sum of the two single limits, if they exist.*

---

[1]we have used the same ε since its choice is arbitrary.

2.7 Limit of a sum or product

Proof

Let $f(x)$ and $g(x)$ be two functions and suppose that they admit limit for $x \to x_0$, i.e.

$$\lim_{x \to x_0} f(x) = l_1,$$

$$\lim_{x \to x_0} g(x) = l_2.$$

Then, with obvious symbology,

$$\forall \varepsilon > 0 \ \exists \delta_1 > 0 : \forall x \in (x_0 - \delta_1, x_0 + \delta_1)$$

$$\Longrightarrow f(x) \in (l_1 - \varepsilon, l_1 + \varepsilon),$$

$$\forall \varepsilon > 0 \ \exists \delta_2 > 0 : \forall x \in (x_0 - \delta_2, x_0 + \delta_2)$$

$$\Longrightarrow g(x) \in (l_2 - \varepsilon, l_2 + \varepsilon).$$

Hence, said $\delta = \min(\delta_1, \delta_2)$,

$$\forall \varepsilon > 0 \ \exists \delta > 0 : \forall x \in (x_0 - \delta, x_0 + \delta),$$

we have

$$l_1 - \varepsilon < f(x) < l_1 + \varepsilon,$$

$$l_2 - \varepsilon < g(x) < l_2 + \varepsilon,$$

from which also

$$l_1 + l_2 - 2\varepsilon < f(x) + g(x) < l_1 + l_2 + 2\varepsilon,$$

which concludes the proof, in fact this last relation, being valid $\forall \varepsilon$, can also be written as

$$l_1 + l_2 - \varepsilon < f(x) + g(x) < l_1 + l_2 + \varepsilon,$$

which is the definition of $l_1 + l_2$ as the limit of $f(x) + g(x)$.

Proposition 2.7.2 (limit of a product). *The limit, for $x \to x_0$, of the product of two functions is equal to the product of the two single limits, if they exist finites.*

Proof

Analogously to the previous theorem we put

$$\lim_{x \to x_0} f(x) = l_1,$$

$$\lim_{x \to x_0} g(x) = l_2,$$

with $|l_1|, |l_2| < \infty$. The conditions

$$\forall \varepsilon > 0 \ \exists \delta_1 > 0 : \forall x \in (x_0 - \delta_1, x_0 + \delta_1)$$

$$\implies f(x) \in (l_1 - \varepsilon, l_1 + \varepsilon),$$

$$\forall \varepsilon > 0 \ \exists \delta_2 > 0 : \forall x \in (x_0 - \delta_2, x_0 + \delta_2)$$

$$\implies g(x) \in (l_2 - \varepsilon, l_2 + \varepsilon),$$

2.7 Limit of a sum or product

can be written as

$$|f(x) - l_1| < \varepsilon, \qquad (2.7.1)$$

$$|g(x) - l_2| < \varepsilon. \qquad (2.7.2)$$

Consider the quantity

$$f(x)g(x) - l_1 l_2,$$

by adding and subtracting the product $l_2 f(x)$ we have

$$f(x)\bigl(g(x) - l_2\bigr) + l_2\bigl(f(x) - l_1\bigr)$$

and, remembering that the modulus of the sum is always less than or equal to the sum of the moduli, we arrive at

$$|f(x)g(x) - l_1 l_2| \le |f(x)||g(x) - l_2| + |l_2||f(x) - l_1|.$$

Using the Eqs. (2.7.1), (2.7.2), we obtain

$$|f(x)g(x) - l_1 l_2| < \bigl(|f(x)| + |l_2|\bigr)\varepsilon$$

which is like writing, being $\forall \varepsilon$ and $|l_2| < \infty$,

$$|f(x)g(x) - l_1 l_2| < \varepsilon$$

and this concludes the proof. Observe that if one of the two limits is $\pm\infty$ and the other is finite but non-zero then the limit for the product is $\pm\infty$.

2.8 Theorem of the permanence of the sign

Theorem 2.8.1 (theorem of the permanence of the sign). *Let $f(x)$ be a function with domain $D = (a,b)$ that has a positive limit for $x \to x_0$, i.e.*

$$\lim_{x \to x_0} f(x) = l > 0.$$

Therefore there exists a neighborhood $\mathfrak{I}$ of x_0 such that

$$\forall x \in \mathfrak{I} \cap D \implies f(x) > 0.$$

Proof

For the definition of limit we have

$$\forall \varepsilon > 0 \ \exists \delta > 0 : \forall x \in (x_0 - \delta, x_0 + \delta)$$

$$\implies f(x) \in (l - \varepsilon, l + \varepsilon),$$

being valid $\forall \varepsilon > 0$ must also be valid for $\varepsilon = l/2$, i.e. there must exist a neighborhood of x_0 with radius $\delta > 0$ such that $\forall x$ in this neighborhood

$$f(x) \in \left(\frac{l}{2}, \frac{3l}{2}\right)$$

and therefore, since $l > 0$, $f(x)$ is also positive and this is true for all values in the neighborhood of radius $\delta > 0$.

2.9 Squeeze theorem

Theorem 2.9.1 (squeeze theorem). *Given three functions $f(x)$, $g(x)$ and $h(x)$ such that, considering a common accumulation point x_0 belonging on the three domains,*

$$\lim_{x \to x_0} g(x) = l,$$

$$\lim_{x \to x_0} h(x) = l,$$

if there exists a neighborhood of x_0 for which

$$g(x) \leq f(x) \leq h(x), \qquad (2.9.1)$$

then we have also

$$\lim_{x \to x_0} f(x) = l.$$

Proof

For the definition of limit, $\forall \varepsilon > 0$ we can write the following relations

$$l - \varepsilon \leq g(x) \leq l + \varepsilon$$

and

$$l - \varepsilon \leq h(x) \leq l + \varepsilon,$$

valid in suitable neighborhoods of x_0.

Furthermore, from Eq. (2.9.1), $\forall \varepsilon > 0$ there exists a neighborhood of x_0, given by the intersection of the neighborhoods in which the above two relations are valid and the neighborhood related to Eq. (2.9.1), where

$$l - \varepsilon \leq g(x) \leq f(x) \leq h(x) \leq l + \varepsilon,$$

$$l - \varepsilon \leq f(x) \leq l + \varepsilon,$$

hence the function $f(x)$ has the limit l for $x \to x_0$.

We briefly summarize some properties of the limits discussed above and others of obvious demonstration in Table 2.9.1, where $f(x)$ and $g(x)$ are two functions such that

$$\lim_{x \to x_0} f(x) = l_1,$$

$$\lim_{x \to x_0} g(x) = l_2.$$

Table 2.9.1

Properties of limits		
function	limit	considerations
$f(x) \pm g(x)$	$l_1 \pm l_2$	
$f(x) \cdot g(x)$	$l_1 \cdot l_2$	$|l_1|, |l_2| < \infty$
$f(x)/g(x)$	l_1/l_2	$|l_1|, |l_2| < \infty, l_2 \neq 0$

2.10 Notable limits

We have the following notable limits:

Proposition 2.10.1 (notable limits 1). *The following relation is an identity*

$$\lim_{x \to 0} \frac{\sin x}{x} = 1.$$

Proof

First of all, the function is even for a neighborhood of 0 of radius $\pi/2$ and therefore we will prove the identity only for $x > 0$. With reference to Figure 2.10.1, for small angles x, we can write the following inequality

$$\sin x \leq x \leq \tan x.$$

Dividing both sides by $\sin x > 0$ and taking their reciprocal we

have
$$\cos x \leq \frac{\sin x}{x} \leq 1.$$
Taking the limit for $x \to 0$, being
$$\lim_{x \to 0} \cos x = 1,$$
by the squeeze theorem we obtain
$$\lim_{x \to 0} \frac{\sin x}{x} = 1.$$

Proposition 2.10.2 (notable limits 2). *The following relation is an identity*
$$\lim_{x \to 0} \frac{1 - \cos x}{x} = 0.$$

Proof

Multiplying and dividing the quantity
$$\frac{1 - \cos x}{x},$$
by $(1 + \cos x)$ we obtain
$$\frac{1 - \cos x}{x} = \frac{\sin^2 x}{x(1 + \cos x)},$$
from which the limit
$$\lim_{x \to 0} \frac{1 - \cos x}{x} = \lim_{x \to 0} \frac{\sin^2 x}{x(1 + \cos x)} = 0.$$

2.10 Notable limits

Figure 2.10.1: *Unit circle where the segments* $\sin x$ *and* $\tan x$ *and the arc of length x are shown with dashed lines.*

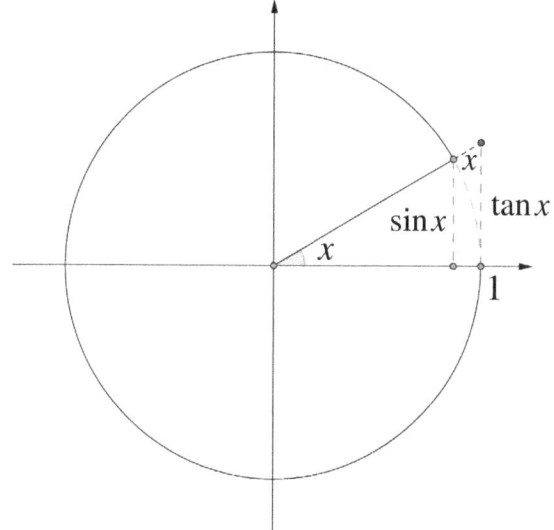

Proposition 2.10.3 (notable limits 3). *The following relation is an identity*

$$\lim_{x \to 0} \frac{1-\cos x}{x^2} = \frac{1}{2}.$$

Proof

Similarly, multiplying and dividing by $(1+\cos x)$ the quantity

$$\frac{1-\cos x}{x^2},$$

we obtain
$$\frac{1-\cos x}{x^2} = \frac{\sin^2 x}{x^2(1+\cos x)}$$
and the limit becomes
$$\lim_{x \to 0} \frac{1-\cos x}{x^2} = \lim_{x \to 0} \frac{\sin^2 x}{x^2(1+\cos x)}$$
$$= \lim_{x \to 0} \left(\frac{\sin x}{x}\right)^2 \cdot \lim_{x \to 0} \frac{1}{1+\cos x} = \frac{1}{2}.$$

Proposition 2.10.4 (notable limits 4). *It can be shown that the following limit*
$$\lim_{x \to \infty} \left(1 + \frac{1}{x}\right)^x = e,$$
is a number between 2 and 3. This number is called the Euler's number and is denoted with e. The Euler's number is irrational and represents the base of natural logarithms.

Proposition 2.10.5 (notable limits 5). *The following relation is an identity*
$$\lim_{x \to 0} \frac{\ln(x+1)}{x} = 1.$$

Proof

Using the properties of logarithms we have
$$\lim_{x \to 0} \frac{\ln(x+1)}{x} = \lim_{x \to 0} \ln(x+1)^{1/x},$$

2.10 Notable limits

Table 2.10.1

Notable limits for $x \to x_0$		
function	x_0	limit
$\frac{\sin x}{x}$	0	1
$\frac{1-\cos x}{x}$	0	0
$\frac{1-\cos x}{x^2}$	0	1/2
$\frac{\ln(x+1)}{x}$	0	1
$\frac{e^x-1}{x}$	0	1
$\left(1+\frac{1}{x}\right)^x$	∞	e

and making the replacement

$$y = 1/x \to x = 1/y,$$

we obtain the equivalent limit

$$\lim_{x \to 0} \ln(x+1)^{1/x} = \lim_{y \to \infty} \ln\left(1+\frac{1}{y}\right)^y = \ln e = 1,$$

where we used the result of the previous notable limit.

Proposition 2.10.6 (notable limits 6). *The following relation is an identity*

$$\lim_{x \to 0} \frac{e^x - 1}{x} = 1.$$

Proof

We can make the replacement

$$y = e^x - 1 \rightarrow x = \ln(y+1)$$

and obtain the equivalent limit

$$\lim_{x \to 0} \frac{e^x - 1}{x} = \lim_{y \to 0} \frac{y}{\ln(y+1)}.$$

Finally, thanks to the value of the previous notable limit we have

$$\lim_{y \to 0} \frac{y}{\ln(y+1)} = 1.$$

In Table 2.10.1 are reported the principal notable limits.

3. Sequences and series

3.1 Introduction

Sequences and series are very useful tools for dealing with various topics, including the concept of integral and discrete systems. Sometimes it can be useful to consider the sum, finite or infinite, of terms of a discrete sequence, their infinite sum is called series.

3.2 Definition of sequence

A sequence can be defined as an infinite sequence of elements each denoted by a discrete index. A sequence, in the abstract sense, is indicated with

$$\{a_n\}_{n \in \mathbb{N}}, \qquad (3.2.1)$$

if the discrete index n varies in the set of natural numbers. Its elements will be denoted by

$$a_1, a_2, a_3, \cdots a_k, \cdots .$$

By writing a_k we mean the k-th element of the sequence shown in Eq. (3.2.1). Examples of sequences are

$$\{a_n\}_n = \left\{\frac{1}{n}\right\}_{n \in \mathbb{N}^+},$$

which defines the sequence of the reciprocals of the natural numbers and

$$\{a_n\}_n = \left\{\frac{1}{n^2}\right\}_{n \in \mathbb{N}^+},$$

which defines, instead, the sequence of the reciprocal squares of the natural numbers. In general, when we want to talk about the sequence $\{a_n\}_{n \in \mathbb{N}}$, we will simply write a_n. In this way a_n could mean both the sequence and its n-th term and it will be necessary to understand from the context which is the correct meaning. Sequences can be considered functions defined on a discrete domain. A sequence can be also defined recursively, an example is the following

$$a_n = \begin{cases} \sqrt{a_{n-1}} + 1, & \text{if } n \geq 1 \\ 2, & \text{if } n = 0, \end{cases}$$

which defines the sequence with the following first terms

$$2, \sqrt{2}+1, \sqrt{\sqrt{2}+1}+1, \cdots.$$

3.3 Limit of sequences

In order to consider the limit of a sequence, it must be observed that the only value to which the index n can tends, being an element of a discrete set, is $+\infty$ in case of $n \in \mathbb{N}$ or $\pm\infty$ in case of $n \in \mathbb{Z}$. In fact these are the only allowed point of accumulation for those sets. Therefore we can write (in the following, unless otherwise specified, we will assume that $n \in \mathbb{N}$)

$$\lim_{n \to +\infty} a_n, \qquad (3.3.1)$$

to indicate the limit value to which the sequence of Eq. (3.2.1) tends when its index tends to[1] $+\infty$. In case the limit shown in Eq. (3.3.1) exists finite, being l its value, we will say that the sequence converges to l. Otherwise in case the limit is infinite we will say that the sequence diverges and, finally, if it does not exist, we will say that the sequence is indeterminate. The following[2] are three example sequences with these three behaviors for $n \to \infty$

$$\frac{1}{n}, \quad \sqrt{n}, \quad (-1)^n.$$

In order, the first converges to zero, the second diverges to $+\infty$, and the third is indeterminate.

3.4 Definition of series

A series is the sum of all terms of a sequence. Given a sequence a_n with $n \in \mathbb{N}$, consider the partial sum S_k defined, for

---

[1] in the case of a sequence with $n \in \mathbb{N}$ we will write simply ∞ meaning $+\infty$.

[2] if not indicated, we assume that $n \in D \subseteq \mathbb{N}$, with D an appropriate domain.

3.4 Definition of series

each index $k \in \mathbb{N}$, as

$$S_k = \sum_{n=0}^{k} a_n, \qquad (3.4.1)$$

we define the series associated to a_n the following

$$\sum_{n=0}^{\infty} a_n := \lim_{k \to \infty} S_k = \lim_{k \to \infty} \sum_{n=0}^{k} a_n.$$

The limit can exist finite, infinite or not exist at all, we will therefore refer to series converging to a certain value, divergent or indeterminate. Later we will show some theorems that relate the behavior of a sequence to that of its series. A series (the same holds also for a sequence) is said to be positive or negative if all the terms are positive or negative, respectively. It is said definitively positive (negative) if the terms are all positive (negative) starting from a certain index. A sequence whose terms alternate in sign is called an alternating sequence.

3.5 Algebraic and geometric sequences

A particular sequence called algebraic sequence, with initial term $a_0 \in \mathbb{R}$ and algebraic reason $h \in \mathbb{R}$, is the following

$$a_n = \begin{cases} h + a_{n-1}, & \text{if } n \geq 1 \\ a_0, & \text{if } n = 0 \end{cases}. \qquad (3.5.1)$$

Another, called geometric sequence, with initial term $b_0 \in \mathbb{R}$ and geometric reason $q \in \mathbb{R}$, is the following

$$b_n = \begin{cases} q \cdot b_{n-1}, & \text{if } n \geq 1 \\ b_0, & \text{if } n = 0 \end{cases}. \qquad (3.5.2)$$

In other words, an algebraic sequence as the property that taking any two consecutive terms, their difference is constant and is called the algebraic reason h. Analogously the ratio of a term of a geometric sequence and its previous one is constant and is called the geometric reason q.

3.5.1 Term n-th

We can find a formula to calculate the k-th term starting from the initial value and the reason of the sequence. This means that it is possible to make explicit the recursive definitions

3.5 Algebraic and geometric sequences

of Eq. (3.5.1) and (3.5.2). The first terms of the algebraic sequence defined in Eq. (3.5.1) are

$$a_0, \quad a_1 = a_0 + h, \quad a_2 = a_1 + h = a_0 + 2h,$$

$$a_3 = a_2 + h = a_0 + 3h, \quad a_n = a_0 + hn.$$

Similarly, the first terms for the geometric sequence defined in Eq. (3.5.2) are

$$b_0, \quad b_1 = b_0 \cdot q, \quad b_2 = b_1 \cdot q = b_0 \cdot q^2,$$

$$b_3 = b_2 \cdot q = b_0 \cdot q^3, \quad b_n = b_0 \cdot q^n.$$

So that, given an algebraic sequence a_n of initial term a_0 and reason h its n-th term is

$$a_n = a_0 + hn, \qquad (3.5.3)$$

while the corresponding formula for a geometric sequence of initial term b_0 and reason q is

$$b_n = b_0 \cdot q^n. \qquad (3.5.4)$$

3.5.2 Partial sum n-th

Similarly to what has been done so far, we can find a formula also for the partial sum n-th. From the definition of Eq. (3.4.1), considering the algebraic sequence defined in Eq. (3.5.1) we have

$$S_n = \sum_{k=0}^{n} a_k = \sum_{k=0}^{n} a_0 + \sum_{k=0}^{n} hk = a_0 \sum_{k=0}^{n} 1 + h \sum_{k=0}^{n} k.$$

The value of the first summation is trivially $(n+1)$, while the second is the sum of the natural numbers from 0 to n and its value is

$$\sum_{k=0}^{n} k = \frac{n(n+1)}{2}.$$

Finally, putting the above formulas together, given an algebraic sequence a_n of initial term a_0 and reason h, for its partial sum n-th we obtain

$$\sum_{k=0}^{n} a_k = (n+1)\left(a_0 + \frac{hn}{2}\right), \qquad (3.5.5)$$

which can be also written as

$$\sum_{k=0}^{n} a_k = \frac{(a_0 + a_n)(n+1)}{2}.$$

For the geometric sequence of Eq. (3.5.2), using the Eq. (3.5.4), we can write the partial sum

$$S_n = \sum_{k=0}^{n} b_k = b_0 \sum_{k=0}^{n} q^k.$$

3.5 Algebraic and geometric sequences

To obtain this result we write

$$b_0 \sum_{k=0}^{n} q^k = b_0(1+q+q^2+q^3+\cdots+q^n)$$

and multiply both members by $(1-q)$

$$b_0(1-q) \sum_{k=0}^{n} q^k = b_0\left((1-q)+(q-q^2)+\cdots+(q^n-q^{n+1})\right).$$

In the second member all the terms are canceled except the first and last so that

$$b_0(1-q) \sum_{k=0}^{n} q^k = b_0(1-q^{n+1}).$$

Finally, given a geometric sequence b_n of initial term b_0 and reason q, its partial sum n-th is

$$\sum_{k=0}^{n} b_k = b_0 \left(\frac{1-q^{n+1}}{1-q} \right), \qquad (3.5.6)$$

which can be also written as

$$\sum_{k=0}^{n} b_k = \frac{b_0 - q b_n}{1-q}.$$

3.5.3 A particular geometric series

The algebraic and geometric series can be calculated by taking the limit $n \to \infty$ of the relations in Eqs. (3.5.5) and (3.5.6).

The geometric series, in particular, plays a very important role for many uses in various disciplines. In particular, consider a geometric series with initial term 1 and reason x, its n-th partial sum, using the Eq. (3.5.6) with $q = x$ and $b_0 = 1$, is

$$\sum_{k=0}^{n} x^k = \frac{1 - x^{n+1}}{1 - x}.$$

This relation is true $\forall x$ and taking the limit for $n \to \infty$ we obtain the following behavior

$$\sum_{k=0}^{\infty} x^k = \begin{cases} \frac{1}{1-x}, & \text{if } |x| < 1 \\ +\infty, & \text{if } x \geq 1 \\ \text{irregular}, & \text{if } x \leq -1 \end{cases},$$

so the series is convergent only for $|x| < 1$. In particular the relation

$$\sum_{k=0}^{\infty} x^k = \frac{1}{1-x} \quad |x| < 1,$$

is very useful and is used in various fields.

3.6 Theorems

We show a basic theorem that gives the necessary condition for a series to converge.

3.6 Theorems

Theorem 3.6.1 (theorem 1). *Given a sequence a_n, if its series converges, i.e.*

$$\left|\sum_{n=0}^{\infty} a_n\right| < \infty,$$

then necessarily

$$\lim_{n \to \infty} a_n = 0.$$

Proof

Consider the partial sum n-th for the sequence a_k defined in Eq. (3.4.1), i.e.

$$S_n = \sum_{k=0}^{n} a_k,$$

if the series converges it means that there exists $l \in \mathbb{R}$ such that

$$\lim_{n \to \infty} S_n = l, \quad |l| < \infty,$$

but also, of course,

$$\lim_{n \to \infty} S_{n-1} = l.$$

The partial sum can be written as

$$S_n = S_{n-1} + a_n,$$

taking the limit for $n \to \infty$ we have

$$l = l + \lim_{n \to \infty} a_n,$$

from which, finally
$$\lim_{n \to \infty} a_n = 0.$$

This theorem is useful because given a series it's quite simple to verify whether the associated sequence converges to zero. If the associated sequence does not converge to zero, thanks to this theorem, we can immediately conclude that the series does not converge. Remember that the condition of convergence to zero of the sequence is a necessary condition, but not sufficient for the convergence of the series.

3.7 Comparison test

Theorem 3.7.1 (comparison test). *Given two sequences a_n and b_n with non-negative terms, such that*
$$a_n \leq b_n,$$
then the following implications hold for the associated series
$$\sum_{n=0}^{\infty} a_n \text{ diverges} \implies \sum_{n=0}^{\infty} b_n \text{ diverges},$$
$$\sum_{n=0}^{\infty} b_n \text{ converges} \implies \sum_{n=0}^{\infty} a_n \text{ converges}.$$

3.7 Comparison test

Proof

Denoting with A_n and B_n the partial sums, i.e.

$$A_n := \sum_{k=0}^{n} a_k,$$

$$B_n := \sum_{k=0}^{n} b_k,$$

we can write immediately, thanks to the hypothesis $a_n \leq b_n$,

$$A_n \leq B_n.$$

Since the sequences are non-decreasing, we have

$$\sup A_n = \lim_{n \to \infty} A_n,$$

$$\sup B_n = \lim_{n \to \infty} B_n,$$

from which

$$\sup A_n \leq \sup B_n.$$

So if the series associated with a_n diverges then

$$\infty \leq \sup B_n, \quad \Longrightarrow \quad \sup B_n = \infty$$

and therefore the series associated with b_n diverges.

Similarly, if the series associated with b_n converges to a finite value L then we can write

$$\sup A_n \leq L$$

and therefore the series associated with a_n converges.

3.8 Asymptotic comparison test

Theorem 3.8.1 (asymptotic comparison test). *Given two sequences a_n and b_n with positive terms such that*

$$\lim_{n \to \infty} \frac{a_n}{b_n} = L \neq 0,$$

then the following implications hold for the associated series

$$\sum_{n=0}^{\infty} a_n \text{ converges} \iff \sum_{n=0}^{\infty} b_n \text{ converges}, \quad (3.8.1)$$

$$\sum_{n=0}^{\infty} a_n \text{ diverges} \iff \sum_{n=0}^{\infty} b_n \text{ diverges}. \quad (3.8.2)$$

Proof

From the definition of limit

$$\forall \varepsilon > 0 \, \exists \tilde{n} : \forall n > \tilde{n} \implies L - \varepsilon \leq \frac{a_n}{b_n} \leq L + \varepsilon,$$

from which, being valid $\forall \varepsilon > 0$ and being a_n and b_n sequences with positive terms,

$$(L-1)b_n \leq a_n \leq (L+1)b_n.$$

Since L is finite by hypothesis, the series associated to the two sequences have the same behavior thanks to the comparison test and we obtain the relations summarized in Eqs. (3.8.1) and (3.8.2).

3.9 Ratio test

Theorem 3.9.1 (criterio del rapporto). *Let a_n be a sequence with positive terms. If*

$$\frac{a_{n+1}}{a_n} \leq L, \qquad (3.9.1)$$

with $L < 1$ then the series converges, if instead

$$\frac{a_{n+1}}{a_n} \geq 1, \qquad (3.9.2)$$

the series diverges.

Proof

Using the Eq. (3.9.1) for the variable n together with the same expression but for the variable $(n+1)$, starting from $n = 0$, we obtain

$$a_n \leq L^n a_0,$$

hence, if $L < 1$, since the geometric series with reason $L < 1$ converges, the series converges thanks to the comparison test.

In the second case we observe that the sequence a_n is positive and not decreasing, therefore

$$\lim_{n \to \infty} a_n \neq 0$$

and the series cannot converge due to the necessary condition on convergence. The series is therefore divergent, being a series of positive terms.

3.10 Asymptotic ratio test

Theorem 3.10.1 (asymptotic ratio test). *Given a sequence a_n with positive terms such that*

$$\lim_{n \to \infty} \frac{a_{n+1}}{a_n} = L,$$

then if $L < 1$ the series converges, if $L > 1$ the series diverges, while in the case $L = 1$ the theorem gives no information.

Proof

In the case $L < 1$ for the definition of limit we can choose $0 < \varepsilon < 1 - L$, for which $\exists \tilde{n} \in \mathbb{N}$ such that $\forall n > \tilde{n}$

$$L - \varepsilon < \frac{a_{n+1}}{a_n} < L + \varepsilon$$

and the series converges thanks to the ratio test, in fact $L+\varepsilon < 1$. In the case $L > 1$ you can choose $\varepsilon < L-1$ in order to have $L-\varepsilon > 1$ and the series diverges again thanks to the ratio test.

3.11 Absolute convergence test

Theorem 3.11.1 (absolute convergence test)**.** *Let a_n be a sequence such that its series converges absolutely, i.e. the series*

$$\sum_{k=0}^{\infty} |a_n|$$

is convergent. Then it also converges

$$\sum_{k=0}^{\infty} a_n.$$

Proof

Let us consider the two sequences

$$b_n = \frac{|a_n| - a_n}{2},$$

$$c_n = \frac{|a_n| + a_n}{2},$$

being $a_n \leq |a_n|$ and observing that

$$0 \leq b_n \leq |a_n|, \qquad 0 \leq c_n \leq |a_n|,$$

from the convergence of the series associated to $|a_n|$, by the comparison test, the series associated with b_n and c_n converge. Observing now that

$$a_n = c_n - b_n,$$

we can conclude that the series associated with the sequence a_n also converges, because it is the difference of two converging series. Note that if a series does not converge absolutely, it does not necessarily mean that it does not converge.

3.12 Root test

We state this theorem without giving a proof.

Theorem 3.12.1 (criterio della radice). *Given a sequence a_n with positive terms for which the limit exists*

$$\lim_{n \to \infty} (a_n)^{1/n} = L,$$

then if $L < 1$ the series converges, if $L > 1$ the series diverges, while in the case $L = 1$ the theorem gives no information.

3.13 Leibniz's test

We formulate now an important theorem for alternate term sequences.

Theorem 3.13.1 (criterio di Leibniz). *Given a sequence a_n with non-increasing positive terms and such that*

$$\lim_{n \to \infty} a_n = 0,$$

then the series

$$\sum_{n=0}^{\infty} (-1)^n a_n$$

converges.

Proof

Let S_n be the n-th partial sum associated with the sequence a_k, defined in Eq. (3.4.1), i.e.

$$S_n = \sum_{k=0}^{n} a_k.$$

The sequence a_n is non-increasing, so we have for every n

$$S_{2n} \geq S_{2n+2} = S_{2n} - a_{2n+1} + a_{2n+2}.$$

Similarly for the partial sums with odd index

$$S_{2n+1} \leq S_{2n+3} = S_{2n+1} + a_{2n+2} - a_{2n+3}.$$

The partial sums with even index is non-increasing, while that with odd index is non-decreasing, moreover

$$S_{2n+1} \leq S_{2n},$$

being

$$S_{2n+1} = S_{2n} - a_{2n+1}.$$

Denoted with A and B the two limits

$$\lim_{n \to \infty} S_{2n+1} = A,$$

$$\lim_{n \to \infty} S_{2n} = B,$$

we have

$$S_{2n+1} \leq A \leq B \leq S_{2n},$$

for all n. Knowing that

$$S_{2n} - S_{2n+1} = a_{2n+1},$$

then, being a_n an infinitesimal sequence, the difference between the sequences of the odd and even partial sums tends to zero, when n diverges. Therefore, by the comparison test we have

$$A = B.$$

3.13 Leibniz's test

Using the definition of limit we have that, for even n,

$$\forall \varepsilon > 0 \,\exists\, m > 0 : |A - S_n| < \varepsilon \quad \forall n > m$$

and, for odd n,

$$\forall \varepsilon > 0 \,\exists\, m' > 0 : |A - S_n| < \varepsilon \quad \forall n > m'$$

then, called $k = \max\{m, m'\}$, we obtain that the relation $|A - S_n|$ holds for every $n > k$, hence, finally,

$$\lim_{n \to \infty} a_n = A.$$

4. Derivatives

4.1 Incremental ratio and derivative

Definition 4.1.1 (derivative). Given a function

$$f: \mathcal{D} \subseteq \mathbb{R} \longrightarrow \mathbb{R}, \quad x \longmapsto f(x),$$

we define the incremental ratio of increment h, denoted by $r(x,h)$, the following quantity

$$r(x,h) = \frac{f(x+h) - f(x)}{h}.$$

Geometrically this represents the angular coefficient of the straight line passing through the two points

$$A \equiv (x, f(x)), \quad B \equiv (x+h, f(x+h)).$$

If for a certain $x \in \mathcal{D}$ exists finite the limit

$$f'(x) := \lim_{h \to 0} r(x,h) = \lim_{h \to 0} \frac{f(x+h) - f(x)}{h}, \quad (4.1.1)$$

then this limit is called the first derivative of the function $f(x)$ in x and the function is said to be differentiable at that point. The function $f(x)$ is said to be differentiable in an interval $\mathcal{I} = (a,b)$ if it is differentiable $\forall x \in \mathcal{I}$ and, in this case, $f'(x)$ is called the derivative function of $f(x)$ with domain $\mathcal{I}$. The value assumed by the derived function for a certain x is the angular coefficient of the tangent line to $f(x)$ at its abscissa point x. A completely analogous way to define the derivative of $f(x)$, in a point x_0, is the following limit

$$f'(x_0) := \lim_{x \to x_0} \frac{f(x) - f(x_0)}{x - x_0}.$$

To indicate the derivative of a function $f(x)$ at a point x where it is differentiable, we usually use the following notations

$$f'(x), \quad \frac{d}{dx} f(x), \quad D_x f(x).$$

Definition 4.1.2 (stationary point). A point x_0 is called a stationary point for a differentiable function $f(x)$ if

$$f'(x_0) = 0.$$

4.2 Properties of the derivative

The derivative acts on functions as a linear application, in fact, given two functions $f(x)$, $g(x)$ and a constant c, it satisfies the properties

$$\frac{d}{dx}(f(x)+g(x)) = \frac{d}{dx}f(x) + \frac{d}{dx}g(x)$$

$$\frac{d}{dx}(c \cdot f(x)) = c\frac{d}{dx}f(x).$$

These two and other properties of the derivative will be demonstrate below. The derivative of the sum of two functions is the sum of the derivatives of the single functions, as can be easily verified directly from the definition of Eq. (4.1.1), in fact,

Chapter 4. Derivatives

given two differentiable functions $f(x)$ and $g(x)$ we can write

$$\begin{aligned}(f(x)+g(x))' &= \lim_{h\to 0}\frac{f(x+h)+g(x+h)-f(x)-g(x)}{h}\\ &= \lim_{h\to 0}\frac{f(x+h)-f(x)}{h}\\ &+ \lim_{h\to 0}\frac{g(x+h)-g(x)}{h}\\ &= f'(x)+g'(x).\end{aligned}$$

The derivative of the product, on the other hand, is not, in general, the same as the product of the derivatives, in fact

$$\begin{aligned}(f(x)\cdot g(x))' &= \lim_{h\to 0}\frac{f(x+h)\cdot g(x+h)-f(x)\cdot g(x)}{h}\\ &= \lim_{h\to 0}\left(\frac{f(x+h)\cdot g(x+h)-f(x)\cdot g(x)}{h}\right.\\ &\left.+ \frac{f(x)\cdot g(x+h)-f(x)\cdot g(x+h)}{h}\right)\\ &= \lim_{h\to 0}\frac{1}{h}(f(x+h)-f(x))\cdot g(x+h)\\ &+ f(x)\cdot(g(x+h)-g(x))\\ &= f'(x)g(x)+f(x)g'(x).\end{aligned}$$

In particular it can be seen that the derivative of the product between a constant and a function is the product of the constant by the derivative of the function.

4.3 Derivatives of elementary functions

From the definition of Eq. (4.1.1) we can calculate the derivative of elementary functions. For example, a function such as

$$f(x) = x,$$

has the following derivative

$$f'(x) = \lim_{h \to 0} \frac{x+h-x}{h} = 1,$$

while function

$$f(x) = x^2,$$

has the derivative

$$f'(x) = \lim_{h \to 0} \frac{x^2 + 2xh + h^2 - x^2}{h} = 2x.$$

In general, to calculate the derivative of a monomial of degree n, proceed as follows. Consider the quantity

$$\mathcal{M}_n(x) = x^n,$$

to calculate its derivative we observe that

$$(x+h)^n = x^n + nx^{n-1}h + \binom{n}{2}x^{n-2}h^2 + \cdots$$

$$+\cdots+\binom{n}{n-2}x^2h^{n-2}+nxh^{n-1}+xh^n,$$

where

$$\binom{n}{k}:=\frac{n!}{k!(n-k)!}$$

is the binomial coefficient and $n!$ indicates the factorial of $n \in \mathbb{N}$, i.e.

$$n! = \begin{cases} 1, & \text{if } n=0 \\ n(n-1)!, & \text{if } n>0 \end{cases}.$$

Using a more compact notation we have

$$(x+h)^n = \sum_{k=0}^{n}\binom{n}{k}x^{n-k}h^k,$$

and the derivative of $\mathcal{M}_n(x)$ can be written as

$$\mathcal{M}'_n(x) = \lim_{h \to 0} \frac{(x+h)^n - x^n}{h},$$

$$\mathcal{M}'_n(x) = \lim_{h \to 0}\left(nx^{n-1} + \sum_{k=2}^{n}\binom{n}{k}x^{n-k}h^{k-1}\right),$$

namely

$$\frac{d}{dx}x^n = nx^{n-1}.$$

In case of a polynomial it is sufficient to derive the individual monomials and use the linearity of the derivative operator.

4.3 Derivatives of elementary functions

Table 4.3.1

\multicolumn{3}{c}{**Derivatives of elementary functions**}
$f(x)$	$\frac{d}{dx}f(x)$	considerations
x^z	zx^{z-1}	$z \in \mathbb{R}, \forall x \in \mathbb{R}$
$\sin x$	$\cos x$	$\forall x \in \mathbb{R}$
$\cos x$	$-\sin x$	$\forall x \in \mathbb{R}$
$\tan x$	$\frac{1}{\cos^2 x}$	$\forall x \in \mathbb{R} \setminus \{\frac{\pi}{2}+k\pi\}, \forall k \in \mathbb{Z}$
$\ln x$	$\frac{1}{x}$	$\forall x \in \mathbb{R} \setminus \{0\}$
e^x	e^x	$\forall x \in \mathbb{R}$

In Table 4.3.1 we report the derivatives of some elementary functions.

Sometimes it can be useful to calculate the *n*-th derivative of a function $f(x)$. We denote with $\hat{D}$ the operator which, acting on $f(x)$, produces its derivative. In Table 4.3.2 are reported some useful results.

Table 4.3.2

$f(x)$	$\hat{D}^n f(x)$	considerations
x^n	$n!$	
x^k	$\frac{k! x^{k-n}}{(k-n)!}$	$k \in \mathbb{N}, k > n$
x^k	0	$k \in \mathbb{N}, k < n$
$\ln x$	$-\frac{(-1)^n (n-1)!}{x^n}$	
e^x	e^x	

4.4 Chain rule

Theorem 4.4.1 (chain rule). *To calculate the derivative of a composite function, we use the so-called chain rule. Given two differentiable functions $f(x)$ and $g(x)$ we have*

$$\frac{d}{dx} f(g(x)) = \frac{df(g)}{dg} \cdot \frac{dg(x)}{dx} = f'(g(x)) g'(x).$$

Proof

Let $w(x)$ be the function defined by

$$w(y) = \frac{f(g(x)+y) - f(g(x))}{y} - f'(g(x)),$$

where

$$y = g(xh+h) - g(x)$$

4.4 Chain rule

and $h > 0$. We have

$$f(g(x)+y) - f(g(x)) = yf'(g(x)) + yw(y) \quad (4.4.1)$$

and, given the differentiability of $f(x)$,

$$\lim_{h \to 0} w(y) = 0. \quad (4.4.2)$$

The derivative of the function $f(g(x))$ can be calculated using the definition of the derivative as the limit of the incremental ratio and using the Eq. (4.4.1), obtaining

$$\begin{aligned}\frac{d}{dx}f(g(x)) &= \lim_{h \to 0} \frac{f(g(x+h)) - f(g(x))}{h} \\ &= \lim_{h \to 0} \frac{f(g(x)+y) - f(g(x))}{h} \\ &= \lim_{h \to 0} \frac{yf'(g(x)) + yw(y)}{h},\end{aligned}$$

or

$$\frac{d}{dx}f(g(x)) = \lim_{h \to 0} \frac{yf'(g(x))}{h} + \lim_{h \to 0} \frac{yw(y)}{h}$$

from which finally, taking the limits and using the expression of Eq. (4.4.2),

$$\frac{d}{dx}f(g(x)) = f'(g(x))g'(x).$$

4.5 Weierstrass theorem

We state the Weierstrass theorem without giving a proof.

Theorem 4.5.1 (Weierstrass theorem). *If a function $f(x)$ is continuous over an interval $[a,b]$, then it assumes maximum and minimum in that interval.*

4.6 Fermat's theorem on stationary points

We state Fermat's theorem on stationary points without giving a proof.

Theorem 4.6.1 (Fermat's theorem on stationary points). *Consider a function $f(x)$ and a point x_0 of its domain where the function has a maximum or a minimum. If $f(x)$ is differentiable in x_0 then*

$$f'(x_0) = 0,$$

i.e. x_0 is a stationary point for $f(x)$.

4.7 Rolle's theorem

Theorem 4.7.1 (Rolle's theorem). *Let $f(x)$ be a continuous function in an interval $[a,b]$ and differentiable in (a,b). If*

$$f(a) = f(b)$$

then $\exists x_0 \in (a,b)$ such that

$$f'(x_0) = 0.$$

Proof

Thanks to the Weierstrass theorem the function admits maximum and minimum, respectively called M and m. Let x_1 and x_2 be the abscissas of these points, i.e.

$$f(x_1) = M, \qquad f(x_2) = m,$$

there are two cases: at least one of x_1 and x_2 is in the open interval (a,b) or neither. In the first case it follows from the Fermat's theorem on stationary points that

$$f'(x_1) = 0$$

and therefore in this case $x_0 = x_1$. In the second case x_1 and x_2 are equal to the extremes a and b. From the hypothesis

$f(a) = f(b)$ it follows then that $m = M$ and the function is constant throughout the interval. Therefore the function has null derivative at all points belonging to $[a,b]$ and this concludes the proof.

4.8 Lagrange's theorem

Theorem 4.8.1 (Lagrange's theorem). *Let $f(x)$ be a continuous function in an interval $[a,b]$ and differentiable in (a,b). Then $\exists x_0 \in (a,b)$ such that*

$$f'(x_0) = \frac{f(b) - f(a)}{b - a}.$$

Proof

Consider the function $h(x)$ defined by

$$h(x) = f(x) - f(a) - \frac{f(b) - f(a)}{b - a}(x - a)$$

with domain $[a,b]$. This function is continuous in $[a,b]$ and differentiable in (a,b) because it is a combination of functions which are by hypothesis continuous in $[a,b]$ and differentiable in (a,b). We can calculate the values $h(a)$ and $h(b)$

$$h(a) = f(a) - f(a) - \frac{f(b) - f(a)}{b - a}(a - a) = 0,$$

$$h(b) = f(b) - f(a) - \frac{f(b)-f(a)}{b-a}(b-a) = 0,$$

hence $h(a) = h(b)$. By applying Rolle's theorem to the function $h(x)$ we can say that there exists $x_0 \in (a,b)$ such that $h'(x_0) = 0$. Therefore

$$0 = h'(x_0) = f'(x_0) - \frac{f(b)-f(a)}{b-a},$$

from which, finally,

$$f'(x_0) = \frac{f(b)-f(a)}{b-a}.$$

This theorem is a special case of Cauchy's theorem which follows.

4.9 Cauchy's theorem

Theorem 4.9.1 (Cauchy's theorem). *Let $f(x)$ and $g(x)$ be two continuous functions in an interval $[a,b]$, differentiable in (a,b) and such that $\forall x \in (a,b)$,*

$$g'(x) \neq 0.$$

Then $\exists x_0 \in (a,b)$ such that

$$\frac{f'(x_0)}{g'(x_0)} = \frac{f(b)-f(a)}{g(b)-g(a)}.$$

Proof

Consider the function $h(x)$ defined by

$$h(x) = f(x) - \frac{f(b)-f(a)}{g(b)-g(a)}g(x)$$

with domain $[a,b]$. It is continuous in $[a,b]$ and differentiable in (a,b) because it is a combination of functions with these properties. Calculating $h(a)$ and $h(b)$ we obtain

$$\begin{aligned} h(a) &= f(a) - \frac{f(b)-f(a)}{g(b)-g(a)}g(a) \\ &= \frac{g(b)f(a)-g(a)f(a)-f(b)g(a)+f(a)g(a)}{g(b)-g(a)} \\ &= \frac{f(a)g(b)-f(b)g(a)}{g(b)-g(a)}, \end{aligned}$$

$$\begin{aligned} h(b) &= f(b) - \frac{f(b)-f(a)}{g(b)-g(a)}g(b) \\ &= \frac{g(b)f(b)-g(a)f(b)-f(b)g(b)+f(a)g(b)}{g(b)-g(a)} \\ &= \frac{f(a)g(b)-f(b)g(a)}{g(b)-g(a)}, \end{aligned}$$

hence $h(a) = h(b)$. By applying Rolle's theorem we can say that there exists $x_0 \in (a,b)$ such that $h'(x_0) = 0$. The derivative of $h(x)$ is

$$h'(x) = f'(x) - \frac{f(b)-f(a)}{g(b)-g(a)}g'(x)$$

and therefore
$$0 = h'(x_0) = f'(x_0) - \frac{f(b)-f(a)}{g(b)-g(a)} g'(x_0),$$
from which, finally,
$$\frac{f'(x_0)}{g'(x_0)} = \frac{f(b)-f(a)}{g(b)-g(a)}.$$
Observe that the hypotheses of the theorem imply that $g(b) \neq g(a)$. In fact, if this were not the case, we could apply Rolle's theorem to $g(x)$ obtaining that in at least one point of the interval (a,b) the derivative $g'(x)$ vanishes in contrast to the hypothesis that $\forall x \in (a,b)$, $g'(x) \neq 0$.

4.10 De L'Hopital's theorem

We state this very useful theorem without giving a proof.

Theorem 4.10.1 (De L'Hopital's theorem). *Let $f(x)$ and $g(x)$ be two continuous functions in an interval $[a,b]$ and differentiable in (a,b) with $g'(x) \neq 0$. If*
$$\lim_{x \to x_0} \frac{f(x)}{g(x)}$$
is an indeterminate form of type $0/0$ or ∞/∞ and if the limit
$$\lim_{x \to x_0} \frac{f'(x)}{g'(x)} = l,$$

exists, then

$$\lim_{x \to x_0} \frac{f(x)}{g(x)} = l.$$

4.11 From mathematics to physics

In physics the concept of derivative is very often used. An example is the definition of the instantaneous velocity of a particle. Suppose that the particle is constrained on a straight line, for example the abscissa axis of a Cartesian reference frame. Its instantaneous velocity is defined as

$$v := \frac{dx(t)}{dt},$$

that is the derivative of its position with respect to time. In the three-dimensional case, the instantaneous velocity is defined, given the position vector $\vec{x} \in \mathbb{R}^3$ of the particle, as

$$\vec{v} := \frac{d\vec{x}(t)}{dt}.$$

Another example can be the concept of potential energy. A conservative force between two particles can be derived from an interaction potential energy V in this way

$$F = -\frac{dV}{dx},$$

4.11 From mathematics to physics

in the one-dimensional case. In the three-dimensional case this is written as[1]

$$\vec{F} = -\vec{\nabla} \cdot V = \left(\frac{\partial V}{\partial x}, \frac{\partial V}{\partial y}, \frac{\partial V}{\partial z}\right),$$

where we have used the vector operator nabla, defined as

$$\vec{\nabla} := \left(\frac{\partial}{\partial x}, \frac{\partial}{\partial y}, \frac{\partial}{\partial z}\right).$$

This operator is used in mathematics and physics and can be called gradient, divergence or rotor depending on whether it is multiplied by a scalar or vector function. The three cases are, respectively,

$$\vec{\nabla} \cdot B, \quad \vec{\nabla} \cdot \vec{A}, \quad \vec{\nabla} \times \vec{A},$$

with $\vec{A}$ vector field and B scalar field. For example, the four Maxwell equations can be written in differential form using the rotor of the electric and magnetic fields and their divergences.

---

[1] the symbol $\frac{\partial}{\partial x}$ indicates the partial derivative with respect to x, i.e. the derivative made only with respect to x of a multi-variable function.

5. Integrals

5.1 Introduction

Consider a continuous function $f(x)$, with domain by D, how can we calculate the area $\mathcal{A}$ that the graph of f delimits with the abscissa axis and two vertical lines? For example, given the function $f(x) = x^2$ you could be interested to know the value of the area subtended by the function between the two vertical lines of equation $x = 0$, $x = 1$ and the abscissa axis. The considered area is shown in Figure 5.1.1.

Chapter 5. Integrals

Figure 5.1.1: *Area subtended by the parabola* $y = x^2$ *between the lines* $x = 0$, $x = 1$ *and the abscissa axis.*

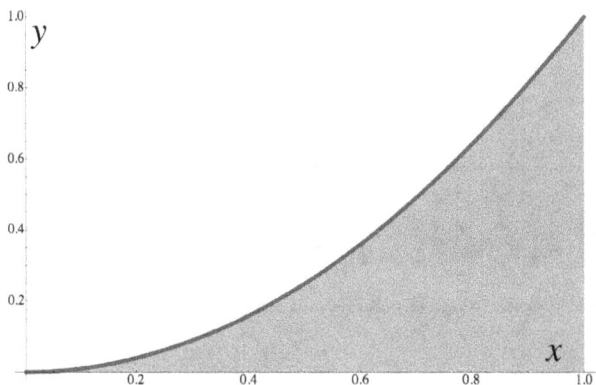

Using the notion of limit, we could divide the interval $I = (0,1)$ on the abscissa axis into n segments of the same length $1/n$. In this way, we obtain $(n+1)$ points on I, whose abscissas are

$$x_k = \frac{k}{n}, \quad k = 0, 1, \cdots, n.$$

We can now calculate approximations for the area A considering the sum of the areas of the rectangles having as base the segments of width $1/n = (x_{k+1} - x_k)$ with $k = 0, 1, \cdots, (n-1)$ and as height of the k-th rectangle the value that the function assumes at one end of the range.

Therefore, by indicating with A^+ and A^- the approximations

5.1 Introduction

under or over the area, we can write

$$\mathcal{A}^+ = \sum_{k=1}^{n} f(x_k) \cdot \frac{1}{n} = \sum_{k=1}^{n} f\left(\frac{k}{n}\right) \cdot \frac{1}{n},$$

$$\mathcal{A}^- = \sum_{k=1}^{n} f(x_{k-1}) \cdot \frac{1}{n} = \sum_{k=1}^{n} f\left(\frac{k-1}{n}\right) \cdot \frac{1}{n},$$

$$\mathcal{A}^- \leq \mathcal{A} \leq \mathcal{A}^+.$$

If

$$\lim_{n \to \infty} \mathcal{A}^- = \lim_{n \to \infty} \mathcal{A}^+,$$

then this value is precisely the desired area. In this case of the parabola $f(x) = x^2$ we have

$$\mathcal{A}^+ = \sum_{k=1}^{n} \left(\frac{k}{n}\right)^2 \cdot \frac{1}{n} = \frac{1}{n^3} \sum_{k=1}^{n} k^2,$$

$$\mathcal{A}^- = \sum_{k=1}^{n} \left(\frac{k-1}{n}\right)^2 \cdot \frac{1}{n} = \frac{1}{n^3} \sum_{k=1}^{n} (k-1)^2,$$

from which

$$\mathcal{A}^+ = \frac{2n^3 + 3n^2 + n}{6n^3},$$

$$\mathcal{A}^- = \frac{2n^3 - 3n^2 + n}{6n^3}$$

and, taking the limit,

$$\lim_{n \to \infty} \mathcal{A}^+ = \lim_{n \to \infty} \mathcal{A}^- = \frac{1}{3},$$

which is the correct result of the subtended area shown in Figure 3.8.2, as we will see later, after the introduction of the concept of integral.

5.2 Definition of integral

Given a generic continuous function and an interval $[a,b]$ on the abscissa axis, consider the n intervals

$$I_k = \left[\frac{(k-1)(b-a)}{n}, \frac{k(b-a)}{n}\right], \quad k = 1, \cdots, n,$$

of width $(b-a)/n$, into which the interval $[a,b]$ is divided. Let x_k be any value in the range I_k, $x_k \in I_k$. The following quantity

$$S_n = \sum_{k=1}^{n} f(x_k) \cdot \frac{b-a}{n},$$

which is an approximation of the area between the function, the abscissa axis and the vertical lines $x = a$ and $x = b$, is called the n-th partial sum.

Definition 5.2.1 (Integrale)**.** We define the (Riemann) integral of the function $f(x)$, in the interval (a,b) and denoted by

$$\int_a^b f(x)\,dx,$$

the limit, if exists finite, for $n \to \infty$ of the partial sums S_n, i.e.

$$\int_a^b f(x)\,dx \equiv \lim_{n \to \infty} S_n.$$

5.3 Linearity of the integral

The integral is a linear operator, i.e. given two continuous functions $f(x)$ and $g(x)$, an interval $[a,b]$ common to the two domains and two constants a,b we have

$$\int_a^b \Big(af(x) + bg(x)\Big)\,dx = a\int_a^b f(x)\,dx + b\int_a^b g(x)\,dx.$$

5.4 Additivity of the integral

The integral is an additive operator, i.e. given a continuous function $f(x)$, a closed interval $[a,b]$ contained in its domain and a point $c \in [a,b]$ we have

$$\int_a^b f(x)\,dx = \int_a^c f(x)\,dx + \int_c^b f(x)\,dx.$$

5.5 Absolute value theorem

We now state this theorem without giving a proof.

Theorem 5.5.1 (absolute value theorem). *Let $f(x)$ be a continuous function and consider a closed interval $[a,b]$ contained in its domain. We have*

$$\left| \int_a^b f(x)\,dx \right| \leq \int_a^b |f(x)|\,dx.$$

5.6 Mean value theorem

Before stating the theorem we give the definition of mean of a function.

Definition 5.6.1 (mean of a function). Let $f(x)$ be a continuous function and let $A = [a,b]$ be a closed interval contained in its domain. The *mean* of the function $f(x)$ on the interval $[a,b]$ is the following

$$\frac{1}{b-a} \int_a^b f(x)\,dx.$$

Intuitively, by associating to the integral between a, b of a positive function the value of the area subtended by its graph, it is possible to associate the ratio between this area and the base $(b-a)$ to the mean of the function (mean height of the geometric figure). In the case of a constant function, the mean is the height of the rectangle that the function subtends.

5.6 Mean value theorem

Theorem 5.6.1 (mean theorem). *Let $f(x)$ be a continuous function defined on a closed interval $D = [a,b]$. We can say that exists a point $x_0 \in D$ for which*

$$\int_a^b f(x)\,dx = f(x_0)(b-a),$$

i.e. the mean of the function is a value that is assumed by the function at least in one point.

Proof

Using Weierstrass theorem there will be a maximum M and a minimum m for the function. We can then write

$$m \leq f(x) \leq M,$$

hence, thanks to the properties of the integral

$$\int_a^b m\,dx \leq \int_a^b f(x)\,dx \leq \int_a^b M\,dx,$$

from which

$$m(b-a) \leq \int_a^b f(x)\,dx \leq M(b-a),$$

which becomes, being $(b-a) > 0$,

$$m \leq \frac{1}{b-a} \int_a^b f(x)\,dx \leq M.$$

Since for a continuous function the image set of an interval contains all the values of the images of each point of the interval then there exists a point $x_0 \in [a,b]$ such that

$$\frac{1}{b-a}\int_a^b f(x) = f(x_0).$$

5.7 Fundamental theorem

The fundamental theorem shows the relationship between the integral and the derivative. Before stating this important theorem it is necessary to give the following

Definition 5.7.1 (primitive). Let $f(x)$ be a function. If exists a differentiable function, denoted with $F(x)$, such that

$$\frac{d}{dx}F(x) = f(x),$$

then the function $F(x)$ is said to be a primitive of $f(x)$. Observe that if a function $f(x)$ admits a primitive $F(x)$ then it admits other infinite primitives: all the functions of the form $F(x) + c$, where c is an arbitrary constant[1]. The set of primitives is called an indefinite integral and is generally indicated

---

[1] in fact $\frac{d}{dx}c = 0$, for all constant c.

5.7 Fundamental theorem

with
$$\int f(x)\,dx = F(x)+c.$$

Note that by definition
$$\frac{d}{dx}\int f(x)\,dx = \frac{d}{dx}(F(x)+c) = f(x).$$

Theorem 5.7.1 (fundamental theorem of integral calculus). *Let $f(x)$ be a continuous function defined on a domain $D = [a,b] \subset \mathbb{R}$ and let $x_0 \in D$, then $\forall \in D$*

$$\frac{d}{dx}\int_a^x f(x)\,dx = f(x), \qquad (5.7.1)$$

furthermore, given a point $b \in D$ and a primitive $F(x)$ of the function $f(x)$, then

$$\int_a^b f(x)\,dx = F(b)-F(a).$$

Proof

Consider the following function
$$g(x) = \int_a^x f(x)\,dx,$$

the incremental ratio has the form
$$\frac{g(x+h)-g(x)}{h} = \frac{1}{h}\left(\int_a^{x+h} f(x)\,dx - \int_a^x f(x)\,dx\right),$$

from which

$$g'(x) = \lim_{h \to 0} \left(\frac{1}{h} \int_x^{x+h} f(x)\,dx \right).$$

Applying the mean theorem there must exist a point $\xi \in [x, x+h]$ such that

$$\frac{1}{h} \int_x^{x+h} f(x)\,dx = f(\xi)$$

and hence

$$g'(x) = \lim_{h \to 0} f(\xi) = f(x),$$

in fact if $h \to 0$ then $\xi \in [x,x], \Longrightarrow \xi = x$.

To demonstrate the second part of the theorem consider

$$g(x) = \int_a^x f(x)\,dx, \qquad (5.7.2)$$

from the proof just made and from the definition of primitive we have

$$g'(x) = f(x) = F'(x),$$

being by hypothesis $F(x)$ a primitive of $f(x)$, the relationship between g and F can only be the following

$$g(x) = F(x) + c, \qquad (5.7.3)$$

with c constant. From Eq. (5.7.2) we have

$$g(b) = \int_a^b f(x)\,dx, \qquad g(a) = \int_a^a f(x)\,dx = 0$$

and therefore we can write

$$\int_a^b f(x)\,dx = g(b) - g(a).$$

Using the Eq. (5.7.3) we obtain

$$\int_a^b f(x)\,dx = F(b) + c - F(a) - c = F(b) - F(a),$$

which concludes the proof.

5.8 Primitives of elementary functions

We report in Table 5.8.1 the primitives of some elementary functions.

Sometimes it may be useful to calculate the n-th primitive with constant zero (the primitive of the primitive and so on n times with $n \in \mathbb{N}$, each time with constant zero) of a continuous function $f(x)$. We denote with $\hat{P}$ the operator which, acting on $f(x)$ produces its primitive with constant zero. We briefly report some useful results, including that of the non-trivial case of $f(x) = \ln x$, in Table 5.8.2.

Primitives of elementary functions

$f(x)$	$\int f(x)\,dx$	considerations
x^z	$\frac{x^{z+1}}{z+1}+c$	$z \in \mathbb{R}\setminus\{-1\}, c \in \mathbb{R}$ constant
$\frac{1}{x}$	$\ln|x|+c$	$c \in \mathbb{R}$ constant
$\sin x$	$-\cos x + c$	$c \in \mathbb{R}$ constant
$\cos x$	$\sin x + c$	$c \in \mathbb{R}$ constant
$\tan x$	$-\ln(\cos x)+c$	$c \in \mathbb{R}$ constant
$\ln x$	$x(\ln x - 1)+c$	$c \in \mathbb{R}$ constant
e^x	$e^x + c$	$c \in \mathbb{R}$ constant

Table 5.8.1

Primitives with $c=0$ of order n, $n \in \mathbb{N}$

$f(x)$	$\hat{P}^n f(x)$	considerations
1	$\frac{x^n}{n!}$	
x^k	$\frac{k!\,x^{k+n}}{(k+n)!}$	$k \in \mathbb{N}^+$
$\ln x$	$\frac{x^n}{n!}\left(\ln x - \sum_{k=1}^{n} 1/k\right)$	
e^x	e^x	

Table 5.8.2

5.9 Methods of integration

Given the close relationship between integral and derivative it is easy to expect that the integral of the product of two functions is not, in general, the product of the two integrals. The relation between these quantities is summarized in the so-called integration by parts.

5.9.1 Integration by parts

To calculate the integral of the product of two continuous functions in an interval $[a,b]$ common to their domains, it is possible to use the following formula

$$\int_a^b f(x)g(x)\,dx = \Big[f(x)G(x)\Big]_a^b - \int_a^b f'(x)G(x)\,dx,$$

where $G(x)$ is a generic primitive of $g(x)$.
To demonstrate this, we can start from

$$\int_a^x f(t)g(t)\,dt = \Big[f(t)G(t)\Big]_a^x - \int_a^x f'(t)G(t)\,dt, \quad (5.9.1)$$

or

$$\int_a^x f(t)g(t)\,dt = f(x)G(x) - f(a)G(a) - \int_a^x f'(t)G(t)\,dt,$$

deriving both members we obtain, from Eq. (5.7.1),

$$f(x)g(x) = \Big(f(x)G(x)\Big)' - f'(x)G(x),$$

$$f(x)g(x) = f'(x)G(x) + f(x)g(x) - f'(x)G(x),$$

which is trivially an identity. Placing $x = b$ in Eq. (5.9.1) we obtain the formula of the integration by parts.

In general we can write

$$\int f(x)g(x)\,dx = f(x)G(x) - \int f'(x)G(x)\,dx,$$

in fact deriving

$$f(x)g(x) = f'(x)G(x) + f(x)g(x) - f'(x)G(x),$$

which is an identity.

5.9.2 Integration by substitution

Another method for calculating integrals is the so-called substitution method. Given the integral

$$\int_a^b f(x)\,dx,$$

we make the substitution

$$x \to y(t),$$

with

$$dx \to \frac{dy(t)}{dt}dt = y'(t)dt$$

and the integral becomes

$$\int_{y^{-1}(a)}^{y^{-1}(b)} f(y(t)) \cdot y'(t) \, dt.$$

5.10 From mathematics to physics

An example of application is the following, suppose we have a string of length L. If the density of the string is not constant at all, but is constant in each circular section of the string, how can you calculate its total mass? We can introduce a linear density $\rho(x)$ which depends on the point x along the string, given a certain origin. The infinitesimal mass dm of an infinitesimal segment of the string with length dx is

$$dm = \rho(x) \, dx.$$

If we consider a reference frame with the origin at one end of the string, we can write

$$m = \int_0^L \rho(x) \, dx,$$

which corresponds to the mass of the string. If $\rho(x)$ is constant, i.e.

$$\rho(x) = \rho,$$

the integral becomes easier to calculate

$$m = \rho \int_0^L dx = \rho \cdot L,$$

as we would expect.

Another example of using integrals is the computation of the flow of a vector physical quantity $\vec{A}$ on a certain closed surface called S. In this case, the required methods of analysis are not covered by this book, but we still want to give a hint of them because the basic idea is intuitive. Indicating with dS the infinitesimal surface element of S we have the flow Φ defined as

$$\Phi = \oint_S dS\, \hat{n} \cdot \vec{A},$$

where $\hat{n}$ is the versor (vector of norm one) perpendicular to dS, with outward direction from dS. The integral is done on a closed surface S and is indicated with the circle on the integral symbol. This kind of integral is of fundamental importance in physics, for example it is used in the integral form of the Maxwell equations in the case where $\vec{A}$ is the electric or magnetic field. Another example is the calculation of the mechanical work in physics, in fact the infinitesimal work dL,

5.10 From mathematics to physics

done on a material point by a force $\vec{F}$ is defined as

$$dL = \vec{F} \cdot d\vec{s},$$

where $d\vec{s}$ is the infinitesimal displacement vector along the path of the material point. Calculating the work means integrating the previous relationship between two points A and B of the trajectory, i.e.

$$L(A \to B) = \int_A^B \vec{F} \cdot d\vec{s},$$

which is a curvilinear integral and generally depends on the path leading from A to B.

6. Exercises

6.1 Exercise 1

The infinitesimal displacement dL along the graph of a differentiable function, associated with an infinitesimal displacement dx along the x-axis, is given by[1]

$$dL = \sqrt{1+(f'(x))^2}\,dx.$$

---
[1] $f'(x)$ is the derivative with respect to x.

Using this expression calculate the length of the graph of the function

$$f(x) = x^{3/2},$$

between the lines $x = 0$ and $x = 4$. In Figure 6.1.1 is shown the length of the curve that should be calculated.

Figure 6.1.1: *Plot of* $f(x) = x^{3/2}$.

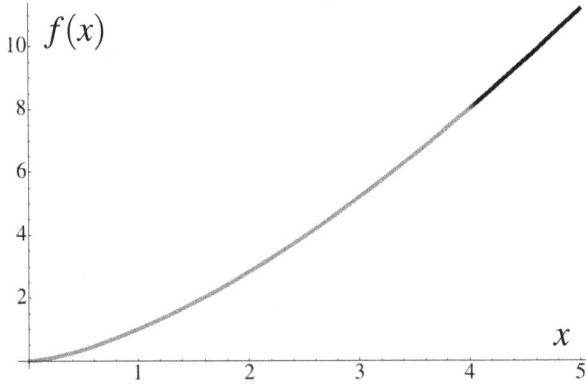

6.2 Exercise 2

Calculate the average value of the function

$$f(x) = xe^{-x^2},$$

between the lines $x = 0$ and $x = 3$. In Figure 6.2.1 is shown the plot of the function.

6.3 Exercise 3

Figure 6.2.1: *Plot of* $f(x) = xe^{-x^2}$.

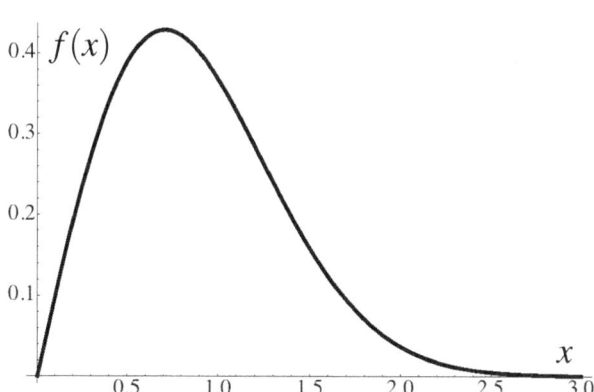

6.3 Exercise 3

Calculate the primitive with zero constant, said $F(x)$, of the function

$$f(x) = \cos\left(\frac{x}{2}\right) + \frac{1}{2}$$

and calculate the area between the functions $F(x)$ and $g(x) = \frac{1}{5}x$, limited by the lines $x = 0$ and $x = 20$. In Figure 6.3.1 is shown area to be calculated.

Figure 6.3.1: *Plot of $F(x)$, primitive with zero constant of $f(x)$, and $g(x) = \frac{1}{5}x$.*

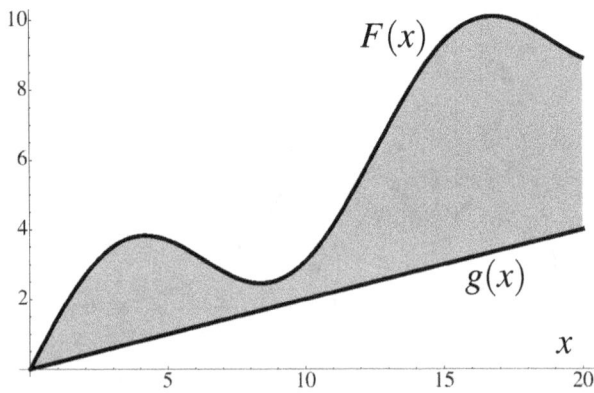

6.4 Exercise 4

Verify that the function

$$\psi(x) = Ae^{ikx} + Be^{-ikx},$$

with A and B constants, is the solution of the Schrödinger stationary equation for free particle, i.e.

$$\frac{d^2}{dx^2}\psi + \frac{2mE}{\hbar^2}\psi = 0,$$

where the modulus of the wave vector, k, is given by $k = \sqrt{\frac{2mE}{\hbar^2}}$ and i is the imaginary unit, with $i^2 = -1$.

6.5 Exercise 5

Let m be the coefficient of the tangent line at the point of abscissa 2 to the graph of the function

$$f(x) = e^{x/4}.$$

Given $A = (5,0)$ and considering P as the point on the graph of the function

$$g(x) = \sqrt{x},$$

that minimizes the distance $\overline{AP}$, determine the equation of the line r passing through P and with angular coefficient m. In Figure 6.5.1 are shown the plots of the two functions $f(x)$ and $g(x)$, together with that of the line r.

Figure 6.5.1: *Plot of $f(x) = e^{x/4}$, $g(x) = \sqrt{x}$ and that of the line r.*

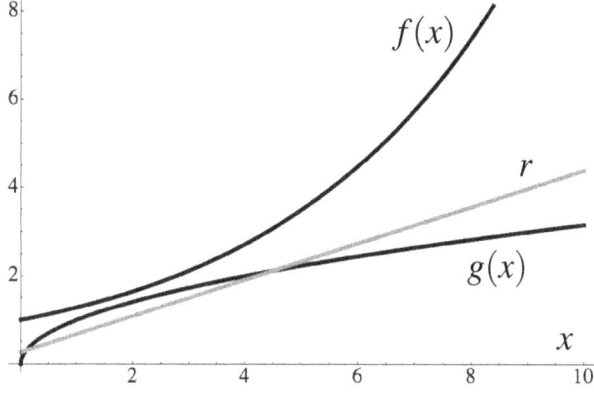

6.6 Exercise 6

Consider a cylinder of base radius r and height h with fixed total surface area (lateral plus bases) with value S. Calculate the base radius under the condition that the volume, said V, is maximum. Indicate, in addition, the value of V. In Figure 6.6.1 there is a cylinder of example.

Figure 6.6.1: *Cylinder of base radius r, height h and total surface area S of the exercise 6.*

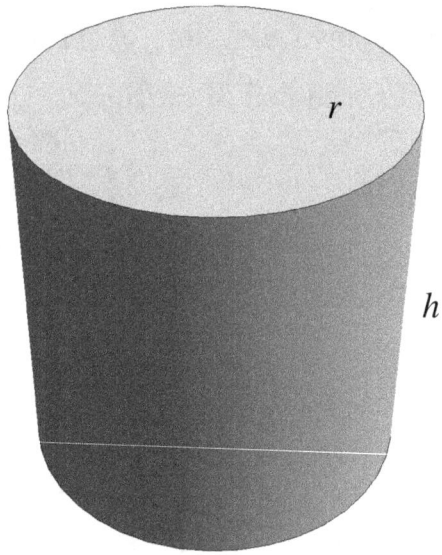

6.7 Exercise 7

Find the constant N, called normalization constant, for the wave function
$$\psi(x) = Nxe^{-x^2},$$
such that[2]
$$\int_{-\infty}^{\infty} \|\psi(x)\|^2 \, dx = 1,$$
i.e. $\psi(x)$ is normalized to 1.

6.8 Exercise 8

Find the position of a classical particle at time $t = 0$, i.e. $x(0)$, knowing that his speed is described by the function
$$v(t) = e^t - \frac{t}{t^2 + 2}$$
and that $x(5) = 200$. Find also the expression $a(t)$ of its acceleration at a general time t. In Figure 6.8.1 is shown a plot for the velocity $v(t)$ for $0 \le t \le 5$.

---

[2] use the Gauss integral: $\int_{-\infty}^{\infty} e^{-x^2} \, dx = \sqrt{\pi}$.

Figure 6.8.1: *Plot of* $v(t) = e^t - \frac{t}{t^2+2}$.

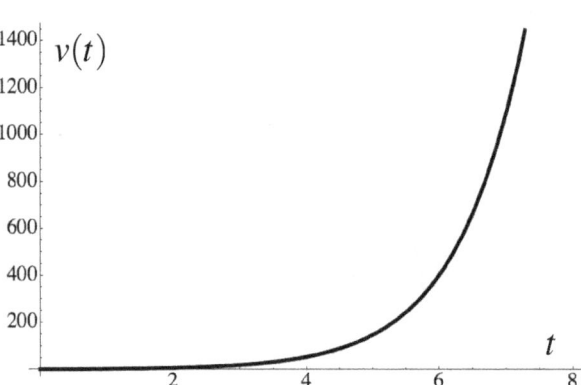

6.9 Exercise 9

Consider a circumference $\mathcal{C}$ of radius r, with center at the origin of a cartesian reference frame. Let $\mathcal{P}$ be the parabola of equation $y = x^2$ and call with P their point of intersection, placed in the first quadrant. Calculate the area A between the parabola, the x-axis, the y-axis and the vertical line s passing through the point P. In Figure 6.9.1 are shown the plots of circumference $\mathcal{C}$, the parabola $\mathcal{P}$ and the line s, in a particular case where $r = 19$.

6.10 Exercise 10

Figure 6.9.1: *Plot of* $\mathcal{C}$, $\mathcal{P}$ *and s of the exercise 9.*

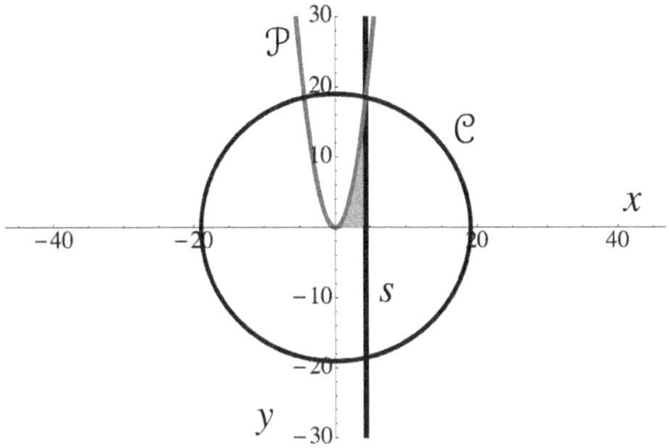

6.10 Exercise 10

Prove by induction that $\forall n \in \mathbb{N}, n \geq 0$ the following equation

$$\sum_{k=0}^{n} k = \frac{n(n+1)}{2}.$$

is an identity.

6.11 Exercise 11

Starting from the relativistic expressions for the total energy of a particle of mass m, i.e.

$$E^2 = m^2 c^4 + \vec{p}^{\,2} c^2$$

and
$$E = T + mc^2,$$
where c is the speed of light in vacuum, T is the kinetic energy and $\vec{p}$ is the momentum of the particle, show that in the non-relativistic limit $T \ll mc^2$, the kinetic energy assumes the expression
$$T = \frac{\vec{p}^2}{2m}.$$

6.12 Exercise 12

Given the function
$$f(x) = kx^2 + \frac{1}{k}x^3,$$
with $k \in (0, 5] \subset \mathbb{R}$, determine the parameter k so that the area under the graph of the function between the lines $x = 0$ and $x = 1$ is minimum.

6.13 Exercise 13

Calculate the following definite integral
$$\int_0^{2\pi} \sin^2 x \, dx.$$
In Figure 6.13.1 is shown the plot of the integrand.

Figure 6.13.1: *Plot of* $f(x) = \sin^2 x$.

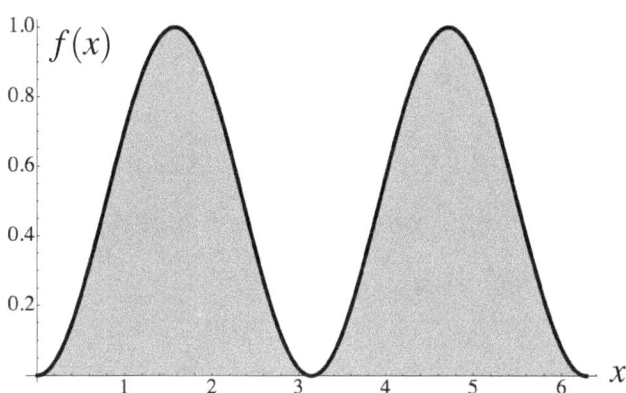

6.14 Exercise 14

Using the definition of derivative of a function demonstrate that
$$\frac{d}{dx} x^n = n x^{n-1}$$
in the case where $n \geq 1$ and $n \in \mathbb{N}$.

6.15 Exercise 15

Consider a third degree polynomial $P(x)$ with

$$P(0) = 1, \quad P'(2) = 2, \quad P''(2) = 3, \quad \int_0^1 P(x)\,dx = 4,$$

where the superscripts indicate the order of derivation. Determine the complete expression of the polynomial.

6.16 Exercise 16

Given the function

$$f(x) = \ln x,$$

find the equation of the line r perpendicular to the graph of the function at the point of abscissa 5 and determine the area shown in Figure 7.16.1.

Figure 6.16.1: *Plot of $f(x) = \ln x$ and of line r of the exercise 16.*

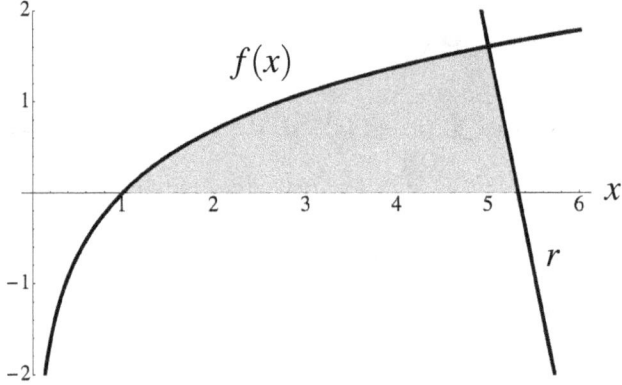

6.17 Exercise 17

Find the radius R of a sphere assuming that the difference between its surface S and its volume V is maximum. What are, in this case, the values of S and V?

6.18 Exercise 18

Calculate the definite integral

$$\int_0^\pi x^3 \sin x\, dx,$$

using a method of your choice. In Figure 6.18.1 is shown the plot of the integrand function.

Figure 6.18.1: *Plot of* $f(x) = x^3 \sin x$.

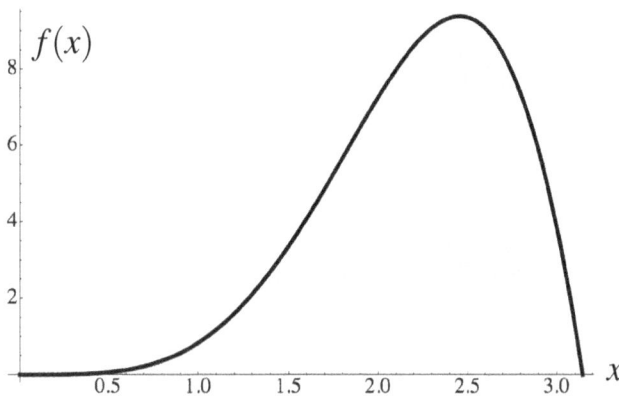

6.19 Exercise 19

Consider the function

$$f(x) = \int_{x-2}^{x+1} \ln(2t - 5)\, dt.$$

Calculate the first derivative of the function, $f'(x)$, and the value of $f'(5)$. In Figure 6.19.1 is shown the plot of $f(x)$ with $0 \leq x \leq 100$.

Figure 6.19.1: *Plot of $f(x)$ of the exercise 19.*

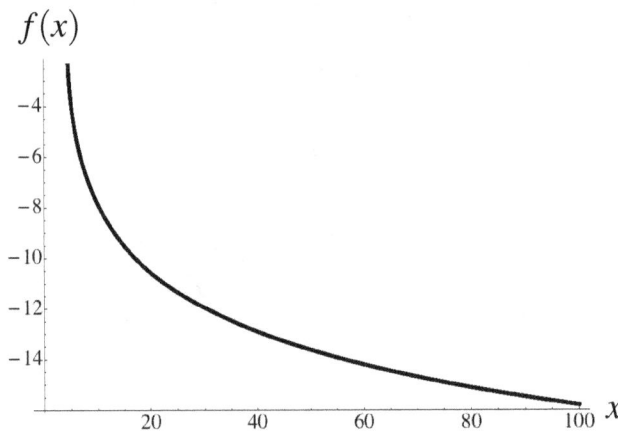

6.20 Exercise 20

Calculate the coefficients a and b so that the relation

$$\frac{1}{x^2 - 5x + 4} = \frac{a}{x - 1} + \frac{b}{x - 4}$$

is an identity. Use the result to calculate the integral

$$\int_5^{10} \frac{1}{x^2 - 5x + 4} dx.$$

In Figure 6.20.1 is shown the plot of the integrand.

6.21 Exercise 21

Figure 6.20.1: *Plot of* $f(x) = \frac{1}{x^2-5x+4}$.

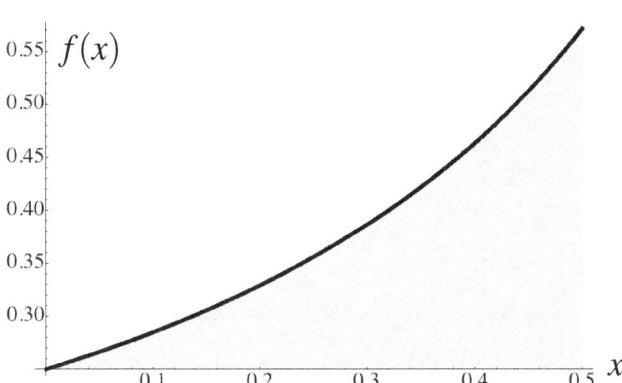

6.21 Exercise 21

Consider, in an appropriate a cartesian reference frame with axes (x,y), the parabola $\mathcal{P}$ of equation

$$y = x^2$$

and the line r of equation

$$y = 3.$$

Consider the rectangles that can be constructed below the line r, having the four vertices as follows: the first in $(0,3)$, the second in the y-axis, the third on the parabola and the fourth on the line r. Find the length of the sides of the rectangle with maximum perimeter. In Figure 6.21.1 is shown the plot of $\mathcal{P}$ and r with a rectangle of example.

Figure 6.21.1: *Plot of the parabola $\mathcal{P}$ of equation $y = x^2$ and of the line r of equation $y = 3$.*

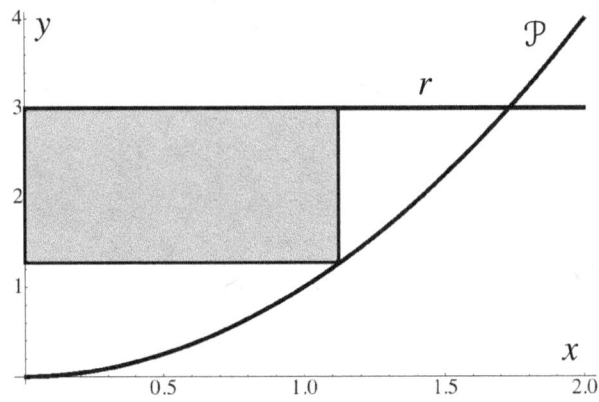

6.22 Exercise 22

Calculate the definite integral

$$\int_{-1}^{1} \sqrt{1-x^2}\, dx.$$

What does it represent geometrically? In Figure 6.22.1 is shown the area represented by the integral.

Figure 6.22.1: *Plot of the function* $f(x) = \sqrt{1-x^2}$.

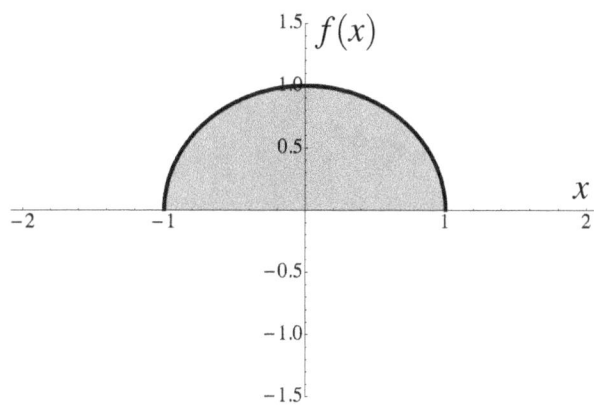

6.23 Exercise 23

Find the equation of the line r tangent to the graph of the function

$$f(x) = xe^{x^2},$$

at the point P of abscissa $\frac{1}{\sqrt{2}}$. In Figure 6.26.1 are shown the plots of $f(x)$ and of the line r.

Figure 6.23.1: *Plot of the function $f(x) = xe^{x^2}$, in black, and that of the line r.*

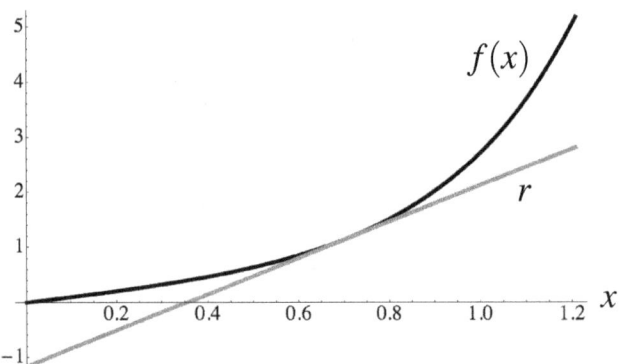

6.24 Exercise 24

Consider the real function

$$f(x) = \ln\left(\frac{\sin x}{1/2 + \cos x}\right).$$

Find the domain of the function, with the limitation $0 \leq x \leq 2\pi$, and calculate its derivative. In Figure 6.24.1 is shown the plot of the function.

Figure 6.24.1: *Plot of the function* $f(x) = \ln\left(\frac{\sin x}{1/2 + \cos x}\right)$ *for* $0 \leq x \leq 2\pi$.

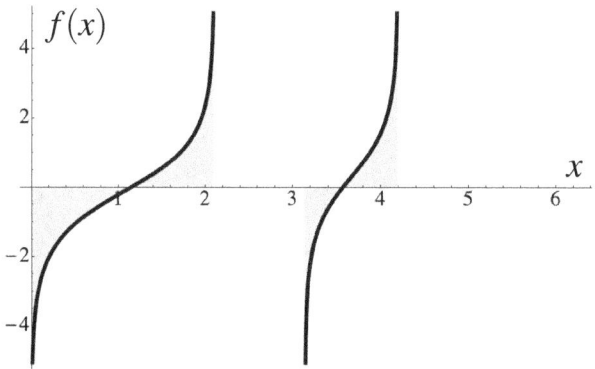

6.25 Exercise 25

The value of a certain property $\mathcal{P}$ at time t is given by

$$\mathcal{P}(t) = e^{-\frac{(t+4)^2}{6}}$$

Calculate the instants where the property $\mathcal{P}$ reaches local maxima and local minima. What happens when $t \to \infty$? In Figure 6.25.1 is shown the plot of $\mathcal{P}(t)$ for $0 \leq t \leq 4$.

Figure 6.25.1: *Plot of the function* $\mathcal{P}(t) = e^{-\frac{(t+4)^2}{6}}$.

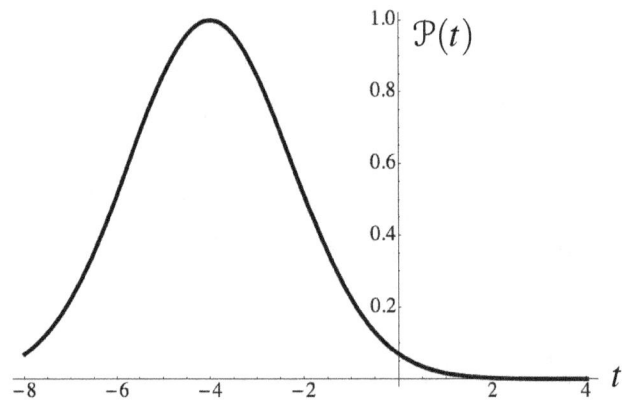

6.26 Exercise 26

Consider, on an appropriate a cartesian reference frame, the function

$$y = \sqrt{x+1}.$$

Find the coordinates of its point P closer to

$$A \equiv (2, 0)$$

and calculate the distance $\overline{PA}$. In Figure 6.26.1 is shown the plot of $y = \sqrt{x+1}$, for $0 \leq t \leq 4$.

6.27 Exercise 27

Figure 6.26.1: *Plot of the function $f(x) = \sqrt{x+1}$ and of the point $A \equiv (2,0)$.*

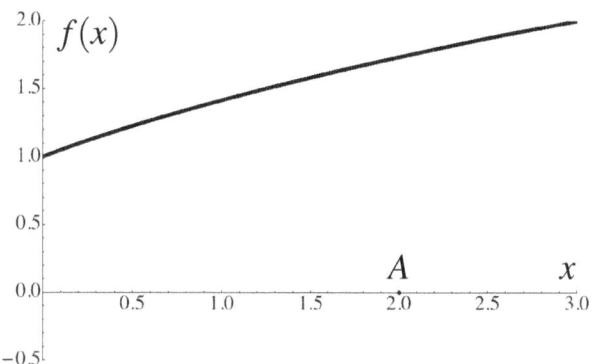

6.27 Exercise 27

Prove, using the method of induction, that $\forall n \in \mathbb{N}, n \geq 1$ the following relation

$$\sum_{k=1}^{n}(2k-1) = n^2$$

is an identity

6.28 Exercise 28

Calculate the following definite integral

$$\int_0^3 x\sqrt{9-x^2}\,dx.$$

6.29 Exercise 29

Find a primitive for the function

$$\sin^2(x)$$

and use it to calculate

$$\int_0^\pi \sin^2(x)\,dx.$$

7. Solutions

7.1 Exercise 1

7.1.1 Text

The infinitesimal displacement dL along the graph of a differentiable function, associated with an infinitesimal displacement dx along the x-axis, is given by[1]

$$dL = \sqrt{1 + (f'(x))^2}\, dx.$$

---
[1] $f'(x)$ is the derivative with respect to x.

Using this expression calculate the length of the graph of the function

$$f(x) = x^{3/2},$$

between the lines $x = 0$ and $x = 4$.

7.1.2 Solution

We should calculate

$$L = \int_0^4 \sqrt{1 + (f'(x))^2}\, dx, \qquad (7.1.1)$$

where

$$f(x) = x^{3/2}.$$

The derivative of the function $f(x)$ is

$$f'(x) = \frac{3}{2} x^{1/2},$$

hence the integral of Eq. (7.1.1) becomes

$$L = \int_0^4 \sqrt{1 + \frac{9}{4} x}\, dx. \qquad (7.1.2)$$

A primitive, said $F(x)$, of the integrand is

$$F(x) = \frac{8}{27}\left(1 + \frac{9}{4} x\right)^{3/2}$$

and we can calculate the value of Eq. (7.1.2) as

$$L = \left[\frac{8}{27}\left(1 + \frac{9}{4} x\right)^{3/2}\right]_0^4 = \frac{8}{27} 10^{3/2} - \frac{8}{27} = \frac{8}{27}(10^{3/2} - 1) \cong 9.1.$$

7.2 Exercise 2

7.2.1 Text

Calculate the average value of the function

$$f(x) = xe^{-x^2},$$

between the lines $x = 0$ and $x = 3$.

7.2.2 Solution

It is sufficient to calculate

$$\mathcal{M} = \frac{1}{3}\int_0^3 xe^{-x^2}\,dx.$$

A primitive of the integrand is

$$-\frac{1}{2}e^{-x^2}$$

and, finally,

$$\mathcal{M} = -\frac{1}{6}\left[e^{-x^2}\right]_0^3 = \frac{1}{6}(1 - e^{-9}) \cong 0.17.$$

7.3 Exercise 3

7.3.1 Text

Calculate the primitive with zero constant, said $F(x)$, of the function

$$f(x) = \cos\left(\frac{x}{2}\right) + \frac{1}{2}$$

and calculate the area between the functions $F(x)$ and $g(x) = \frac{1}{5}x$, limited by the lines $x = 0$ and $x = 20$.

7.3.2 Solution

First of all we can write

$$F(x) = 2\sin\left(\frac{x}{2}\right) + \frac{x}{2}.$$

The area, called $\mathcal{A}$, can be calculated as the difference between the area under $F(x)$ and the area under $\frac{x}{5}$, always between the lines $x = 0$ and $x = 20$. We have to calculate

$$\begin{aligned}
\mathcal{A} &= \int_0^{20} \left(F(x) - \frac{x}{5}\right) dx \\
&= \int_0^{20} \left(2\sin\left(\frac{x}{2}\right) + \frac{x}{2} - \frac{x}{5}\right) dx \\
&= \left[-4\cos\left(\frac{x}{2}\right) + \frac{3x^2}{20}\right]_0^{20}.
\end{aligned}$$

After simplifying the calculations we get

$$\mathcal{A} = 64 - 4\cos 10 \cong 67.4.$$

7.4 Exercise 4

7.4.1 Text

Verify that the function

$$\psi(x) = Ae^{ikx} + Be^{-ikx},$$

with A and B constants, is the solution of the Schrödinger stationary equation for free particle, i.e.

$$\frac{d^2}{dx^2}\psi + \frac{2mE}{\hbar^2}\psi = 0,$$

where the modulus of the wave vector, k, is given by $k = \sqrt{\frac{2mE}{\hbar^2}}$ and i is the imaginary unit, with $i^2 = -1$.

7.4.2 Solution

In order to complete the exercises we can simply put the expression of ψ in the given differential equation, using the property $i^2 = -1$ and the expression

$$k = \sqrt{\frac{2mE}{\hbar^2}}.$$

The second derivative of ψ is

$$\frac{d^2}{dx^2}\psi = (ik)^2 Ae^{ikx} + (-ik)^2 Be^{-ikx},$$

from which

$$\frac{d^2}{dx^2}\psi = -k^2(Ae^{ikx} + Be^{-ikx}) = -k^2\psi$$

and hence

$$\frac{2mE}{\hbar^2}\psi = k^2\psi.$$

Using the given differential equation we have

$$-k^2\psi + k^2\psi = 0$$

and the equation is satisfied.

7.5 Exercise 5

7.5.1 Text

Let m be the coefficient of the tangent line at the point of abscissa 2 to the graph of the function

$$f(x) = e^{x/4}.$$

Given $A = (5,0)$ and considering P as the point on the graph of the function

$$g(x) = \sqrt{x},$$

7.5 Exercise 5

that minimizes the distance $\overline{AP}$, determine the equation of the line r passing through P and with angular coefficient m.

7.5.2 Solution

To obtain m we calculate the derivative of the function $f(x)$, with $x = 2$, i.e.

$$m = \left(\frac{d}{dx}e^{x/4}\right)\bigg|_{x=2} = \left(\frac{1}{4}e^{x/4}\right)\bigg|_{x=2} = \frac{1}{4}\sqrt{e}.$$

A general point P on the graph of the function $g(x)$ can be written as

$$P = (x, \sqrt{x})$$

and its distance D from $A = (5, 0)$ is

$$D = \sqrt{(x-5)^2 + x} = \sqrt{x^2 - 9x + 25}.$$

Minimizing D is analogous to minimize D^2, so we calculate

$$\frac{d}{dx}D^2 = 0$$

and we obtain that $x = \frac{9}{2}$ minimizes D, as can be easily verified observing that

$$\frac{d^2}{dx^2}D\bigg|_{x=9/2} = 2 > 0,$$

hence
$$P = \left(\frac{9}{2}, \frac{3}{\sqrt{2}}\right).$$

A generic line with angular coefficient $m = \frac{1}{4}\sqrt{e}$ has the equation
$$y = \frac{1}{4}\sqrt{e}x + q.$$

In order to obtain q we impose that the point P belongs to the line, i.e.
$$q = \frac{3}{8}(4\sqrt{2} - 3\sqrt{e}).$$

Finally, the equation of the line is
$$y = \frac{1}{4}\sqrt{e}x + \frac{3}{8}(4\sqrt{2} - 3\sqrt{e}).$$

7.6 Exercise 6

7.6.1 Text

Consider a cylinder of base radius r and height h with fixed total surface area (lateral plus bases) with value S. Calculate the base radius under the condition that the volume, said V, is maximum. Indicate, in addition, the value of V.

7.6 Exercise 6

7.6.2 Solution

The total surface S of the cylinder is

$$S = 2\pi rh + 2\pi r^2,$$

that is constant, then one can derive h as a function of r, as

$$h(r) = \frac{S - 2\pi r^2}{2\pi r}.$$

The volume of the cylinder is

$$V = \pi r^2 h$$

and, substituting the expression for h, we obtain

$$V = \frac{Sr}{2} - \pi r^3.$$

To obtain the maximum we calculate $\frac{d}{dr}V = 0$, from which

$$r = \sqrt{\frac{S}{6\pi}}.$$

This value for the radius corresponds to a maximum, in fact

$$\frac{d^2}{dr^2}V\Big|_{r=\sqrt{\frac{S}{6\pi}}} = -6\pi\sqrt{\frac{S}{6\pi}} < 0.$$

The volume is

$$V = \frac{S^{3/2}}{2\sqrt{6\pi}} - \pi\left(\frac{S}{6\pi}\right)^{3/2} = 2\pi\left(\frac{S}{6\pi}\right)^{3/2}.$$

7.7 Exercise 7

7.7.1 Text

Find the constant N, called normalization constant, for the wave function

$$\psi(x) = Nxe^{-x^2},$$

such that[2]

$$\int_{-\infty}^{\infty} \|\psi(x)\|^2 \, dx = 1,$$

i.e. $\psi(x)$ is normalized to 1.

7.7.2 Solution

We have to calculate

$$N^2 \int_{-\infty}^{\infty} x^2 e^{-2x^2} \, dx = 1.$$

We can proceed by integrating by parts, knowing that a primitive, said $F(x)$, for the integrand

$$f(x) = xe^{-2x^2}$$

is

$$F(x) = -\frac{1}{4}e^{-2x^2}.$$

---

[2] use the Gauss integral: $\int_{-\infty}^{\infty} e^{-x^2} \, dx = \sqrt{\pi}$.

7.7 Exercise 7

Therefore we can write

$$\int_{-\infty}^{\infty} x(xe^{-2x^2})\,dx = \left[-\frac{1}{4}xe^{-2x^2}\right]_{-\infty}^{\infty} + \frac{1}{4}\int_{-\infty}^{\infty} e^{-2x^2}\,dx. \quad (7.7.1)$$

The first term in the second member is zero because

$$\lim_{x\to\pm\infty} xe^{-2x^2} = 0,$$

the second term of Eq. (7.7.1) can be calculated by replacing

$$y = \sqrt{2}x, \quad dy = \sqrt{2}dx,$$

and hence

$$\frac{1}{4}\int_{-\infty}^{\infty} e^{-2x^2}\,dx = \frac{1}{4\sqrt{2}}\int_{-\infty}^{\infty} e^{-y^2}\,dy = \frac{\sqrt{\pi}}{4\sqrt{2}}$$

thanks to the Gauss integral. Finally we obtain

$$\int_{-\infty}^{\infty} x^2 e^{-2x^2}\,dx = \frac{\sqrt{\pi}}{4\sqrt{2}}$$

and

$$N = \pm 2\sqrt[4]{\frac{2}{\pi}}$$

7.8 Exercise 8

7.8.1 Text

Find the position of a classical particle at time $t = 0$, i.e. $x(0)$, knowing that his speed is described by the function

$$v(t) = e^t - \frac{t}{t^2+2}$$

and that $x(5) = 200$. Find also the expression $a(t)$ of its acceleration at a general time t.

7.8.2 Solution

Starting from the definition of instantaneous velocity

$$v(t) = \frac{d}{dt}x(t),$$

we can write

$$x(t) = x(0) + \int_0^t v(z)\,dz,$$

from which, in our specific case and for $t = 5$,

$$x(0) = x(5) - \int_0^5 (e^z - \frac{t}{z^2+2})\,dz.$$

A primitive for the integrand function is

$$e^z - \frac{1}{2}\ln(z^2+2),$$

from which, being $x(5) = 200$,

$$x(0) = 200 - \left[e^t - \frac{1}{2}\ln(t^2+2)\right]_0^5,$$

and, finally,

$$x(0) = 200 - e^5 + 1 + \frac{1}{2}\ln\frac{27}{2} \cong 53.9.$$

To find the expression of the acceleration $a(t)$ we can derive $v(t)$, obtaining

$$a(t) = \frac{d}{dt}v(t) = \frac{d}{dt}\left(e^t - \frac{t}{t^2+2}\right) = e^t - \frac{2-t^2}{(t^2+2)^2}$$

7.9 Exercise 9

7.9.1 Text

Consider a circumference $\mathcal{C}$ of radius r, with center at the origin of a cartesian reference frame. Let $\mathcal{P}$ be the parabola of equation $y = x^2$ and call with P their point of intersection, placed in the first quadrant. Calculate the area A between the parabola, the x-axis, the y-axis and the vertical line s passing through the point P.

7.9.2 Solution

To obtain the point P, which is in the first quadrant, we solve the system

$$\begin{cases} y = \sqrt{r^2 - x^2} \\ y = x^2 \end{cases},$$

from which

$$x^4 + x^2 - r^2 = 0.$$

By using $t = x^2$ we can write

$$t^2 + t - r^2 = 0,$$

with the only acceptable solution

$$t = \frac{\sqrt{1 + 4r^2} - 1}{2}.$$

Discarding negative solutions, knowing that $t = x^2$, we can find the abscissa of the point P

$$x_P = \left(\frac{\sqrt{4r^2 + 1} - 1}{2}\right)^{1/2}.$$

The area can be calculated by the definite integral

$$A = \int_0^{x_P} x^2 \, dx = \frac{x_P^3}{3},$$

i.e.

$$A = \frac{1}{3}\left(\frac{\sqrt{4r^2 + 1} - 1}{2}\right)^{3/2}.$$

7.10 Exercise 10

7.10.1 Text

Prove by induction that $\forall n \in \mathbb{N}, n \geq 0$ the following equation

$$\sum_{k=0}^{n} k = \frac{n(n+1)}{2}.$$

is an identity.

7.10.2 Solution

First of all it must be shown that the relation is valid for $n = 0$, so that

$$\sum_{k=0}^{0} k = 0 = \frac{0(0+1)}{2}$$

and it is an identity. It is now sufficient to prove that if the relation of holds for n so it is valid also for $n+1$. We write the relation for $n+1$, i.e.

$$\sum_{k=0}^{n+1} k = \sum_{k=0}^{n} k + (n+1) = \frac{(n+2)(n+1)}{2},$$

from which

$$\sum_{k=0}^{n} k = \frac{(n+2)(n+1) - 2n - 2}{2}.$$

Using for the first member the original relation for n, we obtain
$$\frac{n(n+1)}{2} = \frac{(n+2)(n+1) - 2n - 2}{2},$$
from which
$$\frac{n(n+1)}{2} = \frac{n^2 + 3n + 2 - 2n - 2}{2} = \frac{n^2 + n}{2} = \frac{n(n+1)}{2}$$
and this concludes the proof.

7.11 Exercise 11

7.11.1 Text

Starting from the relativistic expressions for the total energy of a particle of mass m, i.e.
$$E^2 = m^2 c^4 + \vec{p}^{\,2} c^2$$
and
$$E = T + mc^2,$$
where c is the speed of light in vacuum, T is the kinetic energy and $\vec{p}$ is the momentum of the particle, show that in the non-relativistic limit $T \ll mc^2$, the kinetic energy assumes the expression
$$T = \frac{\vec{p}^{\,2}}{2m}.$$

7.11.2 Solution

We can calculate E^2 using the second expression

$$E^2 = T^2 + m^2 c^4 + 2mc^2 T.$$

Using this result in the first expression we obtain

$$\vec{p}^{\,2} c^2 = T^2 + 2mc^2 T,$$

from which, finally,

$$T(T + 2mc^2) = \vec{p}^{\,2} c^2.$$

Neglecting the kinetic energy T in parenthesis with respect to $2mc^2$ for the given condition, we arrive to

$$T = \frac{\vec{p}^{\,2}}{2m}.$$

7.12 Exercise 12

7.12.1 Text

Given the function

$$f(x) = kx^2 + \frac{1}{k}x^3,$$

with $k \in (0, 5] \subset \mathbb{R}$, determine the parameter k so that the area under the graph of the function between the lines $x = 0$ and $x = 1$ is minimum.

7.12.2 Solution

The area, said A, can be obtained by integrating the function between $x = 0$ and $x = 1$. In particular we have

$$A = \int_0^1 \left(kx^2 + \frac{1}{k}x^3\right) dx = k\int_0^1 x^2\, dx + \frac{1}{k}\int_0^1 x^3\, dx.$$

We calculate

$$k\int_0^1 x^2\, dx = \frac{k}{3},$$

from which

$$\frac{1}{k}\int_0^1 x^3\, dx = \frac{1}{4k}.$$

The area is given by

$$A = \frac{k}{3} + \frac{1}{4k} = \frac{4k^2 + 3}{12k}.$$

In order to find the extremes of the function we calculate

$$\frac{d}{dk}A = \frac{1}{3} - \frac{1}{4k^2} = 0, \quad \longrightarrow \quad k = \pm\frac{\sqrt{3}}{2}$$

and we can accept only the following solution

$$k = \frac{\sqrt{3}}{2}$$

due to the domain of variation of k. The second derivative of A is

$$\frac{d^2}{dk^2}A = \frac{1}{2k^3}$$

and, being positive for the chosen k, the latter represents a local minimum. We must check also the right extreme of the interval $(0,5]$, by calculating

$$A(5) = \frac{103}{60} \cong 1.72,$$

to be compared with

$$A\left(k = \frac{\sqrt{3}}{2}\right) = \frac{1}{\sqrt{3}} \cong 0.58.$$

Finally, the value of k that minimize A is

$$k = \frac{\sqrt{3}}{2}.$$

7.13 Exercise 13

7.13.1 Text

Calculate the following definite integral

$$\int_0^{2\pi} \sin^2 x\, dx.$$

7.13.2 Solution

Integrating by parts we can write

$$\int_0^{2\pi} \sin^2 x\, dx = \int_0^{2\pi} \sin x \sin x\, dx$$

$$= [-\cos x \sin x]_0^{2\pi} + \int_0^{2\pi} \cos^2 x\, dx.$$

Using the fundamental relation

$$\cos^2 x = 1 - \sin^2 x,$$

we obtain

$$\int_0^{2\pi} \sin^2 x \, dx = \left[-\cos x \sin x\right]_0^{2\pi} + \int_0^{2\pi} dx - \int_0^{2\pi} \sin^2 x \, dx. \tag{7.13.1}$$

Moreover

$$\left[-\cos x \sin x\right]_0^{2\pi} = 0$$

and

$$\int_0^{2\pi} dx = 2\pi,$$

hence the Eq. (7.13.1) gives

$$\int_0^{2\pi} \sin^2 x \, dx = \frac{1}{2}(2\pi) = \pi.$$

7.14 Exercise 14

7.14.1 Text

Using the definition of derivative of a function demonstrate that

$$\frac{d}{dx} x^n = n x^{n-1}$$

in the case where $n \geq 1$ and $n \in \mathbb{N}$.

7.14 Exercise 14

7.14.2 Solution

To find the derivative of the function we should calculate the limit

$$\frac{d}{dx}x^n = \lim_{h \to 0} \frac{(x+h)^n - x^n}{h} \qquad (7.14.1)$$

We can write[3]

$$(x+h)^n = \sum_{k=0}^{n} \binom{n}{k} x^{n-k} h^k = \sum_{k=1}^{n} \binom{n}{k} x^{n-k} h^k + x^n$$

and, by inserting in the Eq. (7.14.1), we have

$$\frac{d}{dx}x^n = \lim_{h \to 0} \sum_{k=1}^{n} \binom{n}{k} x^{n-k} h^{k-1}$$

$$= \lim_{h \to 0} \left[\sum_{k=2}^{n} \binom{n}{k} x^{n-k} h^{k-1} + \binom{n}{1} x^{n-1} \right].$$

Taking the limit we observe that the first addend goes to zero as it contains the powers of h, obtaining

$$\frac{d}{dx}x^n = \binom{n}{1} x^{n-1} = n x^{n-1},$$

that concludes the proof.

---

[3] $\binom{n}{k} = \frac{n!}{k!(n-k)!}$ is the binomial coefficient.

7.15 Exercise 15

7.15.1 Text

Consider a third degree polynomial $P(x)$ with

$$P(0) = 1, \quad P'(2) = 2, \quad P''(2) = 3, \quad \int_0^1 P(x)\,dx = 4,$$

where the superscripts indicate the order of derivation. Determine the complete expression of the polynomial.

7.15.2 Solution

The form of a generic third degree complete polynomial is

$$P(x) = ax^3 + bx^2 + cx + d.$$

In this case a, b, c, d are constant coefficients to be found to solve the exercise.

From the first condition, $P(0) = 0$, we obtain trivially $d = 1$. The polynomial becomes

$$P(x) = ax^3 + bx^2 + cx + 1. \qquad (7.15.1)$$

The first and second derivatives of the polynomial are

$$P'(x) = 3ax^2 + 2bx + c,$$

$$P''(x) = 6ax + 2b.$$

By using the second and third conditions we can write

$$\begin{cases} 2 = P'(2) = 12a + 4b + c \\ 3 = P''(2) = 12a + 2b \end{cases},$$

which gives

$$c = 12a - 4, \quad b = \frac{3}{2} - 6a.$$

The polynomial of Eq. (7.15.1) becomes now

$$P(x) = ax^3 + \left(\frac{3}{2} - 6a\right)x^2 + (12a - 4)x + 1. \quad (7.15.2)$$

In order to obtain the value of a, we use the last condition, i.e.

$$\begin{aligned} 4 &= \int_0^1 P(x)\,dx \\ &= \int_0^1 \left(ax^3 + \left(\frac{3}{2} - 6a\right)x^2 + (12a - 4)x + 1\right)dx \\ &= \left[\frac{a}{4}x^4 + \left(\frac{1}{2} - 2a\right)x^3 + (6a - 2)x^2 + x\right]_0^1 \\ &= \frac{1}{4}a + \frac{1}{2} + 4a - 1 \\ &= \frac{17}{4}a - \frac{1}{2}, \end{aligned}$$

from which, finally,
$$a = \frac{18}{17}.$$
The polynomial of Eq. (7.15.2) becomes
$$P(x) = \frac{18}{17}x^3 - \frac{165}{34}x^2 + \frac{148}{17}x + 1.$$

7.16 Exercise 16

7.16.1 Text

Given the function
$$f(x) = \ln x,$$
find the equation of the line r perpendicular to the graph of the function at the point of abscissa 5 and determine the area shown in Figure 7.16.1.

7.16.2 Solution

The first derivative of the function at $x = 5$ determines the angular coefficient m of the line tangent to its graph at that point. To obtain the angular coefficient of the perpendicular line we have to calculate the opposite of the reciprocal of m. We have to calculate
$$m = -\frac{1}{f'(5)}.$$

7.16 Exercise 16

Figure 7.16.1: *Plot of $f(x) = \ln x$ and of line r of the exercise 16.*

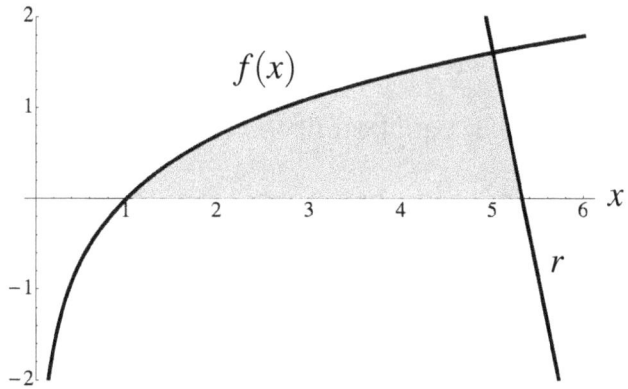

The derivative of the function $f(x)$ is

$$f'(x) = \frac{d}{dx}\ln x = \frac{1}{x},$$

from which

$$m = -\frac{1}{1/5} = -5.$$

A generic line with angular coefficient $m = -5$ has the form

$$y = -5x + q$$

and we need to calculate the value of q by imposing the passage from the point P

$$P \equiv (5, \ln 5).$$

We calculate

$$\ln 5 = -5 \cdot 5 + q \quad \longrightarrow \quad q = \ln 5 + 25,$$

hence the cartesian equation of the line r is

$$y = -5x + 25 + \ln 5.$$

In order to calculate the area, called $\mathcal{A}$, we calculate the abscissa x_A of the point A of intersection between the line r and the x-axis

$$0 = -5x_A + 25 + \ln 5 \quad \longrightarrow \quad x_A = 5 + \frac{\ln 5}{5} \quad (7.16.1)$$

and then we proceed with the definite integral

$$\begin{aligned}
\mathcal{A} &= \int_1^5 \ln x\, dx + \int_5^{x_A} (-5x + 25 + \ln 5)\, dx \\
&= \left[\frac{1}{x}\right]_1^5 + \left[-\frac{5}{2}x^2 + 25x + x\ln 5\right]_5^{x_A} \\
&= \frac{1}{5} - 1 - \frac{5}{2}x_A^2 + (25 + \ln 5)x_A + \frac{125}{2} - 125 - 5\ln 5 \\
&= -\frac{633}{10} - 5\ln 5 - \frac{5}{2}x_A^2 + (25 + \ln 5)x_A.
\end{aligned}$$

Using the value of x_A from Eq. (7.16.1), we obtain

$$\mathcal{A} = -\frac{633}{10} - 5\ln 5 - \frac{5}{2}x_A^2 + (25 + \ln 5)x_A$$

7.17 Exercise 17

7.17.1 Text

Find the radius R of a sphere assuming that the difference between its surface S and its volume V is maximum. What are, in this case, the values of S and V?

7.17.2 Solution

Let $\mathcal{D}$ be the difference between the surface and the volume of the sphere, i.e.

$$\mathcal{D}(r) = S(r) - V(r) = 4\pi r^2 - \frac{4}{3}\pi r^3.$$

To search the maximum we can put to zero the following derivative

$$0 = \frac{d}{dr}\mathcal{D}(r) = 8\pi r - 4\pi r^2,$$

from which $r = 2$. We calculate the second derivative

$$\frac{d^2}{dr^2}\mathcal{D}(r)\bigg|_{r=2} = 8\pi - 16\pi = -8\pi < 0$$

and, being negative, for $r = 2$ the function $\mathcal{D}$ has a local maximum. We observe that from the condition $0 \leq r < \infty$ and

$$\mathcal{D}(0) < \mathcal{D}(2),$$

we can conclude that for $r=2$ the maximum is global. Finally, the values of S and V, with $r=2$, are

$$S(2) = 16\pi, \quad V(2) = \frac{32}{3}\pi.$$

7.18 Exercise 18

7.18.1 Text

Calculate the definite integral

$$\int_0^\pi x^3 \sin x \, dx,$$

using a method of your choice.

7.18.2 Solution

We can obtain the result by integrating by parts several times in this way

$$\begin{aligned}\int_0^\pi x^3 \sin x \, dx &= -\left[x^3 \cos x\right]_0^\pi + 3\int_0^\pi x^2 \cos x \, dx \\ &= \pi^3 + 3\left[x^2 \sin x\right]_0^\pi - 6\int_0^\pi x \sin x \, dx \\ &= \pi^3 - 6\left[x \cos x\right]_0^\pi + 6\int_0^\pi \cos x \, dx.\end{aligned}$$

From which, finally,

$$\int_0^\pi x^3 \sin x \, dx = \pi^3 + 6\pi.$$

7.19 Exercise 19

7.19.1 Text

Consider the function

$$f(x) = \int_{x-2}^{x+1} \ln(2t-5)\, dt.$$

Calculate the first derivative of the function, $f'(x)$, and the value of $f'(5)$.

7.19.2 Solution

We have, in general

$$\frac{d}{dx} \int_{g(x)}^{h(x)} q(t)\, dt = h'(x)q(h(x)) - g'(x)q(g(x)).$$

In this case

$$h(x) = x+1, \quad g(x) = x-2, \quad q(t) = \ln(2t-5),$$

hence

$$f'(x) = \ln(2x+2-5) - \ln(2x-4-5) = \ln\left(\frac{2x-3}{2x-9}\right)$$

and

$$f'(5) = \ln 7 \cong 1.95.$$

7.20 Exercise 20

7.20.1 Text

Calculate the coefficients a and b so that the relation

$$\frac{1}{x^2 - 5x + 4} = \frac{a}{x-1} + \frac{b}{x-4}$$

is an identity. Use the result to calculate the integral

$$\int_5^{10} \frac{1}{x^2 - 5x + 4} dx.$$

7.20.2 Solution

To find the coefficients we write

$$\frac{1}{x^2 - 5x + 4} = \frac{a}{x-1} + \frac{b}{x-4}$$
$$= \frac{ax - 4a + bx - b}{(x-1)(x-4)}$$
$$= \frac{(a+b)x - 4a - b}{x^2 - 5x + 4},$$

from which

$$1 = (a+b)x - 4a - b.$$

In order to have an identity (i.e. an expression valid $\forall x$) it is necessary that the coefficients of the powers of x of the same

7.20 Exercise 20

order are equal in both members. Therefore we have to solve the system

$$\begin{cases} a+b=0 \\ -4a-b=1 \end{cases}$$

from which

$$a=-\frac{1}{3}, \quad b=\frac{1}{3}.$$

We can write

$$\frac{1}{x^2-5x+4} = \frac{1}{3(1-x)} + \frac{1}{3(x-4)}$$

and the integral becomes

$$\begin{aligned}
\int_5^{10} \frac{1}{x^2-5x+4}\,dx &= -\frac{1}{3}\int_5^{10}\frac{1}{x-1}\,dx + \frac{1}{3}\int_5^{10}\frac{1}{x-4}\,dx \\
&= -\frac{1}{3}\Big[\ln(x-1)\Big]_5^{10} + \frac{1}{3}\Big[\ln(x-4)\Big]_5^{10} \\
&= \frac{1}{3}(\ln 4 - \ln 9 + \ln 6 - \ln 1) = \frac{1}{3}\ln\frac{8}{3}.
\end{aligned}$$

Finally we obtain

$$\int_5^{10} \frac{1}{x^2-5x+4}\,dx = \frac{1}{3}\ln\frac{8}{3} \cong 0.33.$$

7.21 Exercise 21

7.21.1 Text

Consider, in an appropriate a cartesian reference frame with axes (x,y), the parabola $\mathcal{P}$ of equation

$$y = x^2$$

and the line r of equation

$$y = 3.$$

Consider the rectangles that can be constructed below the line r, having the four vertices as follows: the first in $(0,3)$, the second in the y-axis, the third on the parabola and the fourth on the line r. Find the length of the sides of the rectangle with maximum perimeter.

7.21.2 Solution

Let

$$A(x,x^2), \quad 0 \le x \le \sqrt{3},$$

be the point of the vertex of the rectangle which is on the parabola, then the sides of the rectangle, l and L can be written

as
$$l = x, \quad L = 3 - x^2.$$

The perimeter is
$$P(x) = 2x + 6 - 2x^2,$$

and we should find a maximum. In this case the point is the vertex of the parabola $P(x)$, as can be seen by calculating the value of the following derivative

$$0 = \frac{d}{dx}(2x + 6 - 2x^2) = 2 - 4x \quad \longrightarrow \quad x = \frac{1}{2}.$$

This is the abscissa of a maximum, because

$$\frac{d^2}{dx^2}(2x + 6 - 2x^2)\bigg|_{x=\frac{1}{2}} = -4 < 0.$$

The perimeter for this local maximum is

$$P\left(\frac{1}{2}\right) = \frac{13}{2},$$

and it should be compared with the value at the extreme

$$P(0) = 6 < P\left(\frac{1}{2}\right), \quad P(\sqrt{3}) = 2\sqrt{3} < P\left(\frac{1}{2}\right).$$

Finally, the sides of the rectangle with maximum perimeter are

$$l = \frac{1}{2}, \quad L = \frac{11}{4}.$$

7.22 Exercise 22

7.22.1 Text

Calculate the definite integral

$$\int_{-1}^{1} \sqrt{1-x^2}\,dx.$$

What does it represent geometrically?

7.22.2 Solution

We can change the variable of integration by replacing

$$x = \sin t, \qquad dx = \cos(t)\,dt$$

so that the integral becomes

$$\int_{-1}^{1} \sqrt{1-x^2}\,dx = \int_{-\frac{\pi}{2}}^{\frac{\pi}{2}} \sqrt{1-\sin^2(t)}\cos(t)\,dt = \int_{-\frac{\pi}{2}}^{\frac{\pi}{2}} \cos^2(t)\,dt,$$

Integrating by parts

$$\int_{-\frac{\pi}{2}}^{\frac{\pi}{2}} \cos^2(t)\,dt = \left[\sin(t)\cos(t)\right]_{-\frac{\pi}{2}}^{\frac{\pi}{2}} + \int_{-\frac{\pi}{2}}^{\frac{\pi}{2}} \sin^2(t)\,dt$$

$$= \pi - \int_{-\frac{\pi}{2}}^{\frac{\pi}{2}} \cos^2(t)\,dt,$$

from which

$$\int_{-\frac{\pi}{2}}^{\frac{\pi}{2}} \cos^2(t)\,dt = \frac{\pi}{2}.$$

Finally
$$\int_{-1}^{1} \sqrt{1-x^2}\,dx = \frac{\pi}{2}$$
and geometrically it represents the area of a semicircle of radius 1.

7.23 Exercise 23

7.23.1 Text

Find the equation of the line r tangent to the graph of the function
$$f(x) = xe^{x^2},$$
at the point P of abscissa $\frac{1}{\sqrt{2}}$.

7.23.2 Solution

The point P is
$$P \equiv \left(\frac{1}{\sqrt{2}}, \sqrt{\frac{e}{2}}\right),$$
as can be seen by replacing the given value $x = 1/\sqrt{2}$ in the function $f(x)$. We consider a line of type
$$y = mx + q \tag{7.23.1}$$

and calculate m using the derivative

$$m = f'(\frac{1}{\sqrt{2}}) = (1+2x^2)e^{x^2}\Big|_{x=\frac{1}{\sqrt{2}}} = 2\sqrt{e}.$$

Imposing the passage of the line of Eq. (7.23.1) through P we find

$$\sqrt{\frac{e}{2}} = 2\sqrt{e}\frac{1}{\sqrt{2}} + q, \quad \longrightarrow \quad q = -\sqrt{\frac{e}{2}},$$

hence the equation of the line r is

$$y = 2\sqrt{e}x - \sqrt{\frac{e}{2}}.$$

7.24 Exercise 24

7.24.1 Text

Consider the real function

$$f(x) = \ln\left(\frac{\sin x}{1/2 + \cos x}\right).$$

Find the domain of the function, with the limitation $0 \le x \le 2\pi$, and calculate its derivative.

7.24.2 Solution

The domain is given by

$$\begin{cases} \frac{1}{2}+\cos x \neq 0 \\ \frac{\sin x}{1/2+\cos x} > 0 \end{cases},$$

Making a study of the signs

	0	$\frac{2}{3}\pi$	π	$\frac{4}{3}\pi$	2π
$\sin x > 0$		+	+	−	−
$1/2 + \cos x > 0$		+	−	−	+
$\frac{\sin x}{1/2+\cos x} > 0$		+	−	+	−

we obtain

$$D = \left(0, \frac{2}{3}\pi\right) \cup \left(\pi, \frac{4}{3}\pi\right).$$

Finally, the derivative is

$$\begin{aligned} f'(x) &= \frac{1/2+\cos x}{\sin x} \cdot \frac{d}{dx}\left(\frac{\sin x}{1/2+\cos x}\right) \\ &= \frac{1/2+\cos x}{\sin x} \cdot \frac{(1/2+\cos x)\cos x + \sin^2 x}{(1/2+\cos x)^2} \\ &= \frac{2+\cos x}{(1+2\cos x)\sin x}. \end{aligned}$$

7.25 Exercise 25

7.25.1 Text

The value of a certain property $\mathcal{P}$ at time t is given by

$$\mathcal{P}(t) = e^{-\frac{(t+4)^2}{6}}$$

Calculate the instants where the property $\mathcal{P}$ reaches local maxima and local minima. What happens when $t \to \infty$?

7.25.2 Solution

We put the first derivative equal to zero

$$0 = \frac{d}{dt}\mathcal{P}(t) = \frac{d}{dt}\left(e^{-\frac{(t+4)^2}{6}}\right) = -\frac{t+4}{3}e^{-\frac{(t+4)^2}{6}},$$

obtaining $t = -4$. This is a local maximum, in fact

$$\frac{d^2}{dt^2}\mathcal{P}(t)\bigg|_{t=-4} = \left(-\frac{1}{3} + \frac{(t+4)^2}{9}\right)e^{-\frac{(t+4)^2}{6}}\bigg|_{t=-4} = -\frac{1}{3} < 0.$$

In addition, we can observe that there are no local minima. For $t \to \infty$ we have

$$\lim_{t \to \infty} e^{-\frac{(t+4)^2}{6}} = \lim_{t \to \infty} e^{-\frac{t^2}{6}} = 0.$$

7.26 Exercise 26

7.26.1 Text

Consider, on an appropriate a cartesian reference frame, the function

$$y = \sqrt{x+1}.$$

Find the coordinates of its point P closer to

$$A \equiv (2,0)$$

and calculate the distance $\overline{PA}$.

7.26.2 Solution

The coordinates of the point P can be written as

$$P \equiv \left(x, \sqrt{x+1}\right).$$

we can calculate the square of the distance D between A and P as follows

$$\overline{PA}^2 = D^2(x) = (x-2)^2 + x + 1 = x^2 - 3x + 5. \quad (7.26.1)$$

Minimize the distance is equivalent to minimizing its square and then we calculate the first derivative of Eq. (7.26.1) and we put it equal to zero,

$$0 = \frac{d}{dx}D^2(x) = 2x - 3, \quad \longrightarrow \quad x = \frac{3}{2}.$$

For $x = \frac{3}{2}$ there is a local minimum, in fact

$$\frac{d^2}{dx^2}D^2(x) = 2 > 0,$$

since the function of Eq. (7.26.1) is a parabola with the concavity upwards. Finally, the distance $\overline{PA}$ has the value

$$\overline{PA} = \sqrt{x^2 - 3x + 5}\Big|_{x=\frac{3}{2}} = \frac{\sqrt{11}}{2} \cong 1.66.$$

7.27 Exercise 27

7.27.1 Text

Prove, using the method of induction, that $\forall n \in \mathbb{N}, n \geq 1$ the following relation

$$\sum_{k=1}^{n}(2k-1) = n^2$$

is an identity

7.27.2 Solution

We verify that the relation is valid for $n = 1$

$$\sum_{k=1}^{1}(2k-1) = 1 = 1^2.$$

Let us prove that if the relation is valid for n so it holds also for $n+1$. For $n+1$ it becomes

$$\sum_{k=1}^{n+1}(2k-1) = \sum_{k=1}^{n}(2k-1)+2n+1 = (n+1)^2,$$

from which

$$\sum_{k=1}^{n}(2k-1) = n^2+1+2n-2n-1 = n^2,$$

which is trivially the relation for n and this concludes the proof.

7.28 Exercise 28

7.28.1 Text

Calculate the following definite integral

$$\int_0^3 x\sqrt{9-x^2}\,dx.$$

7.28.2 Solution

We can use the substitution method by placing

$$x = 3\sin t$$

and hence

$$dx = 3\cos t\,dt.$$

The integral becomes

$$\int_0^3 x\sqrt{9-x^2}\,dx = 9 \int_0^{\pi/2} \sin t \cos t \sqrt{9 - 9\sin^2 t}\,dt,$$

from which

$$\int_0^3 x\sqrt{9-x^2}\,dx = 27 \int_0^{\pi/2} \sin t \cos^2 t\,dt.$$

The latter can be integrated by parts

$$\int_0^{\pi/2} \sin t \cos^2 t\,dt = -\left[\cos^3 t\right]_0^{\pi/2} - 2\int_0^{\pi/2} \cos^2 t \sin t\,dt,$$

which leads to

$$\int_0^{\pi/2} \sin t \cos^2 t\,dt = -\frac{1}{3}\left[\cos^3 t\right]_0^{\pi/2} = \frac{1}{3}.$$

Finally, the integral is

$$\int_0^3 x\sqrt{9-x^2}\,dx = 27 \cdot \frac{1}{3} = 9.$$

7.29 Exercise 29

7.29.1 Text

Find a primitive for the function

$$\sin^2(x)$$

and use it to calculate

$$\int_0^\pi \sin^2(x)\,dx.$$

7.29.2 Solution

This integral can be calculated by parts observing that

$$\int \sin^2 x\, dx = \int \sin x \cdot \sin x\, dx.$$

Therefore

$$\int \sin^2 x\, dx = -\sin x \cos x + \int \cos^2 x\, dx,$$

using the relation

$$\cos^2 x = 1 - \sin^2 x,$$

we arrive to the expression

$$\int \sin^2 x\, dx = -\sin x \cos x + \int 1\, dx - \int \sin^2 x\, dx,$$

or

$$\int \sin^2 x\, dx = -\sin x \cos x + x - \int \sin^2 x\, dx.$$

Finally

$$\int \sin^2 x\, dx = \frac{x - \sin x \cos x}{2} + c$$

with c constant is the set of primitives for the function $\sin^2 x$.
The definite integral is

$$\int_0^\pi \sin^2 x\, dx = \left[\frac{x - \sin x \cos x}{2} \right]_0^\pi = \frac{\pi}{2} \cong 1.57.$$

Other books

Fundamentals of physics

ISBN: 9798655711945

Concepts of physics
ISBN: 9798675717668

Complex numbers
ISBN: 9798674312185

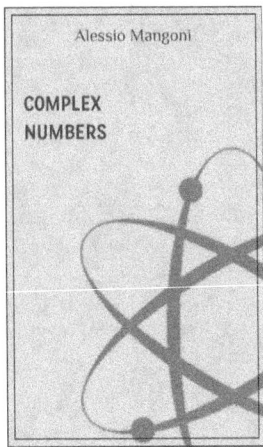

Special relativity

ISBN: 9798675703647

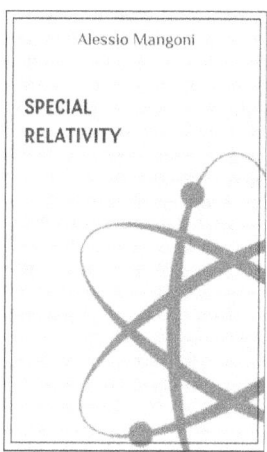

The mathematics of quantum mechanics

ISBN: 9798645275037

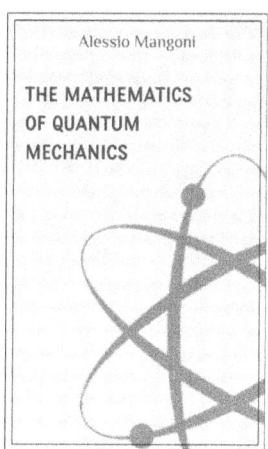

The Dirac equation

ISBN: 9798666724644

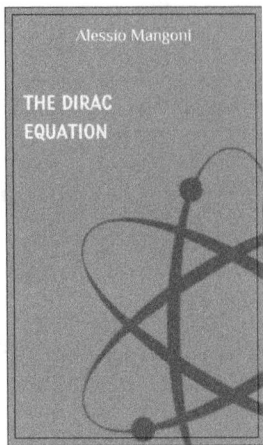

Relativity, decays and electromagnetic fields

ISBN: 9798663840200

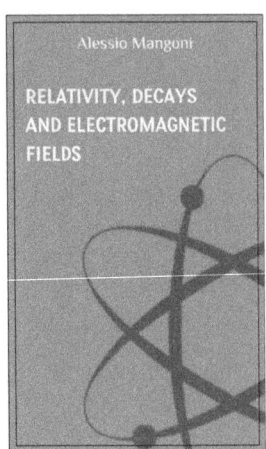

www.ingramcontent.com/pod-product-compliance
Lightning Source LLC
Chambersburg PA
CBHW081427220526
45466CB00008B/2294